Energy, Climate and the Environment

The aim of this series is to provide texts which lay out the technical, environmental and political issues relating to proposed policies for responding to climate change. The focus is not primarily on the science of climate change, or on the technological detail, although there will be accounts of this, to aid assessment of the viability of various options. However, the main focus is the policy conflicts over which strategy to pursue. The series adopts a critical approach and attempts to identify flaws in emerging policies, propositions and assertions. In particular, it seeks to illuminate counter-intuitive assessments, conclusions and new perspectives. The intention is not simply to map the debates, but to explore their structure, their underlying assumptions and their limitations. The books in this series are incisive and authoritative sources of critical analysis and commentary, clearly indicating the divergent views that have emerged whilst also identifying the shortcomings of such views. The series does not simply provide an overview, but also offers policy prescriptions.

Geoffrey Wood · Juan Felipe Neira-Castro
Editors

From Fossil Fuels to Low Carbon Energy Transition

New Regulatory Trends in Latin America

palgrave
macmillan

Editors
Geoffrey Wood
School of Law
University of Stirling
Stirling, UK

Juan Felipe Neira-Castro
Department of Mining and Energy Law
Externado University
Bogota, Colombia

ISSN 2947-8561 ISSN 2947-857X (electronic)
Energy, Climate and the Environment
ISBN 978-3-031-00301-1 ISBN 978-3-031-00299-1 (eBook)
https://doi.org/10.1007/978-3-031-00299-1

Cover illustration: Artur Debat

This Palgrave Macmillan imprint is published by the registered company Springer Nature Switzerland AG
The registered company address is: Gewerbestrasse 11, 6330 Cham, Switzerland

In memory of Ivan Giedelmann Reyes who told me what Latin America is really all about, over endless cups of tinto whilst watching Assassins. Rest in peace, mi hermano *(Dr. Geoffrey Wood)*

Foreword

Energy transition has featured prominently in Latin America policy discussions over the past decade and will continue to be at the centre of the policy debate over the next 30 years, as the world continues to push for carbon neutrality by 2050. However, there a several factors that complicate the discussion in the region given the wide range of policy objectives, natural endowments and political differences that exist today among Latin American countries.

On the one hand, there are countries that have updated their policy and regulatory frameworks to mobilize private resources and accelerate the transition from fossil fuels to a lower carbon economy, with a particular emphasis on the power and transportation sectors. This first group of countries includes, among others, Chile, Colombia, Costa Rica, and Uruguay, which have developed sound regulation and public policies, in many cases including investment-based tax incentives and competitive auctions for long-term power purchase agreements. The results are quite promising, as the share of variable or non-conventional renewable energy has increased significantly in their respective power matrixes, reaching medium-term targets much earlier than originally anticipated.

Policy action in these countries have transcended national borders as Chile, Colombia and Costa Rica co-lead the *RELAC Initiative*, which seeks to increase the collective share of renewable energy sources (both conventional and non-conventional) in the region's power matrix from 56% in 2020 to 70% by 2030. This is the most ambitious regional clean electricity initiative in the world today. Equally important has been the design and implementation of policies to promote sustainable mobility, both for private vehicles and mass transportation systems. The Colombian case is a good example. The country sanctioned the National Electric Vehicle Law in 2019, and since then it has become the regional leader in electric vehicles sales exhibiting double digit growth rates in 2020 and 2021 despite the pandemic. Finally, some of the countries within this group still depend on extractive industries from a macroeconomic perspective. However, there is also very good progress in terms of energy transition and decarbonization efforts within these industries. Examples include the plans to use low-carbon hydrogen or synthetic fuels for mining trucks in Chile and Ecopetrol's, Colombia's national oil company, massive deployment of solar photovoltaic projects to power its petroleum production sites, as well as its plans for wind power and low-carbon hydrogen projects at its refineries.

On the other hand, there are countries that have reversed previous successful clean energy policies in favour of fossil fuels in the name of energy security, such as Mexico, or others in which the policy debate on energy transition is muted, such as Venezuela. Many issues in this group of countries are driven more by politics than policy. The question here is how long will these countries continue to be absent from the global efforts to fight climate change. Finally, there are countries that may fall somewhere in between the two groups of countries described above. In some cases, political forces may undermine existing policies to move towards a lower carbon economy, while in others administration changes may bring a renewed interest in energy transition and climate action.

A cross cutting issue among energy transition and extractive industries is the so-called social licensed to operate and how to resolve disputes with communities, often ethnic ones, in regions where power, petroleum and mining projects are developed. These issues could be compounded if energy transition initiatives also entail labour reconversion processes in places that largely depend on fossil fuel industries. A lesson learned from energy transitions in other jurisdictions shows that these processes need not only be just and well-managed but also people centred.

Many of the policy and regulatory issues that have shaped the energy transition in Latin America are discussed in detail by renowned academics and policy practitioners from the region and elsewhere in *New Regulatory Trends in Latin America: From Fossil Fuels to Low Carbon Energy Transition.* The book, which is divided into two main parts that cover hydrocarbons, electricity and mining, and social license to operate and dispute resolution mechanisms, includes specific case studies that shed light on how countries in the region have dealt with their energy transition process. The book will be a great tool for students, professors, policy makers, and anyone that wishes to be well-informed on energy and extractive industries policy and practice in Latin America.

Diego Mesa Puyo
Minister of Mines and Energy
Government of Colombia
Bogotá, Colombia

The original online version of this Front Matter was revised: The "Foreword" page has been included.

Acknowledgements

We would like to take this opportunity to sincerely thank all those who contributed the excellent chapters to this book. Without all your knowledge, expertise and time, this book would never have happened. We would also like to thank Rachael Ballard, Punitha Balasubramaniam and Prof. Dave Elliott at Palgrave Macmillan, and to all those who gave their support in other ways. Thanks also to Academic Proofreading and Editing Services (https://academicproofreadingandeditingservices.com/) for providing help with the proofreading and editing and to Ian Moore for his useful comments and advice on the manuscript.

Contents

Editors and Contributors

About the Editors

Dr Wood Geoffrey is a Lecturer in Energy and Environmental Law at Stirling University Law School in Scotland. He holds a Ph.D. in low carbon energy governance from the Centre for Energy Petroleum and Mineral Law and Policy (CEPMLP) at the University of Dundee. An interdisciplinary energy and environmental law, policy and regulatory specialist, his research and teaching focuses on the development of legal and governance frameworks for sustainable energy transitions to optimise policy delivery in terms of environmental, social, economic, political and technological outcomes. Recent research includes exploring the impact of austerity on environmental decision making in Scotland, a critical analysis of policy risk and politics on low carbon energy deployment in the UK and Scotland, the role of definitions in renewable and low carbon energy governance. He has published extensively on energy and environmental issues, and has published over 40articles, books, book chapters and other publications, including two books *The Palgrave Handbook of Managing Fossil Fuels and Energy Transitions* (2019) and

A Critical review of Scottish Renewable and Low Carbon Energy Policy (2017). He is also Series Editor of the major new Palgrave Studies in Energy Transitions (PSET) handbook series (Palgrave 2022), Co-Series Editor for the *Energy, Climate and Environment* book series (Palgrave), and Editor-in-Chief for the *Encyclopedia of Renewable and Low Carbon Energy* (forthcoming, Palgrave). Currently, also a Researcher in global energy transitions at the University of Dundee and a Guest Lecturer in energy, climate and environment at Glasgow Caledonian University, he has previously held research, lecturing and consultancy positions for various organisations including the Scottish Government, the Offshore Renewables Institute, Royal Society of Edinburgh and the Extractives Hub (a joint UK Government-University of Dundee project).

Dr Neira-Castro Juan Felipe works at the National Hydrocarbons Agency (ANH), the administrator of the hydrocarbon resources of Colombia based in Bogotá. Previously, he was a Lecturer in Sustainable Development and Energy Policy at the Mining and Energy Law Department at Externado University, Bogotá. He holds an LLM in Petroleum Law and Policy, an M.Sc. in International Oil and Gas Management and a Ph.D. in Energy Law from the Centre for Energy, Petroleum and Mineral Law and Policy (CEPMLP), all from the University of Dundee, Scotland. He has published extensively in areas such as gas regulation in Colombia, unconventional hydrocarbons and energy transition. He has advised international and government entities on public policy and regulatory affairs.

Contributors

Abaunza Daniela Aguilar has a Ph.D. in Law from the University of Waikato (New Zealand), Master in Energy Law and Policy from the University of Dundee (Scotland), Specialist in Energy and Mining Law and Lawyer from University Externado of Colombia, and currently works as a Lecturer and Researcher at the University Externado of Colombia.

Anchustegui Ignacio Herrera is a Latin American (Venezuelan) established in Scandinavia. He is an Associate Professor at the Faculty of Law of the University of Bergen where he leads the Research group for Natural Resource Law, Environmental Law and Development Law. He is also a member of the Bergen Offshore Wind Centre (BOW) and the Bergen Center for Competition Law and Economics (BECCLE). His research focuses on the regulation of energy markets, at the intersection between energy, competition and public procurement law, areas in which he has published extensively, with more than 40 peer-reviewed articles and book contributions. His PhD 'Buyer Power in EU Competition Law' (Concurrences 2017) received the Concurrences 2017 Award for best PhD of the year in Competition Law and Economics. He also received the Meltzer Award for Young Researchers in 2018.

Asencio Andrés Zamorano (Chile), lawyer, Central University. He graduated with a Diploma in Energy and Natural Resources at the P. Universidad Católica de Chile and a LLM in Mineral Law and Policy from the Center for Energy, Petroleum, Mineral Law and Policy (CEPMLP), University of Dundee. His area of study and research is energy law and policy, with emphasis on energy policy and energy security. His practice has been developed in the areas of public service and private areas and international organisations.

Bastidas Reyes Luis Fernando has a Law degree, a Specialization in Administrative Law and a Masters in Economic Law from Pontificia Universidad Javeriana, Bogotá, Colombia, and a Master of Laws, LL.M., from the University of Richmond School of Law, Richmond, VA, USA. He has approximately 11 years of experience in public and corporate executive positions in managerial, regulatory, legal affairs and social license matters. Also, he has strong leadership and teamwork skills and is proficient in the design, implementation and management of policies and projects.

Bustos Luis is Associate Professor at the Externado University in Colombia and an Attorney at Law from the Universidad Externado. He holds the degrees LL.M. in Mineral Law and Policy (Dundee, UK) and a Specialization in Public Law, Political Science, and Sociology (Externado University, Colombia). Currently, he is a Ph.D. student at the Universidad Panamericana de México.

Carrasco Abel M. Venero holds a LL.B. degree awarded by *Pontificia Universidad Católica del Perú* and a LL.M. in Energy Law and Policy degree awarded by the CEPMLP of the University of Dundee. Currently, he is Senior Associate at the Lima office of Santivanez Abogados, a specialised energy law boutique.

de Medeiros Costa Hirdan Katarina is a lawyer, consultant and professor. She has been working in the energy sector since 2004. She has a bachelor's degree in Law, and her undergrad background is Petroleum Law from the University of Rio Grande do Norte (Brazil). At that time, she wrote her monograph on joint operating agreements in the Brazilian oil upstream sector where she explored the international model of contracts in the Brazilian reality after the market opened for foreign investors in 1997. Also, she has published papers in events, conferences books and international journals. Moreover, she has a Master's Degree in Energy Issues from the University of Sao Paulo and her thesis focused on the Regulation of Natural Gas. She has Ph.D. on Energy Issues and her thesis focused on the Oil Brazilian royalties' system. She is Post-Doctorate in Sustainability from the School of Arts, Sciences and Humanities of the University of São Paulo (EACH/USP), Post-Doctorate in Energy from the Graduate Program in Energy of the University of São Paulo (PPGE/USP), and Master of Law from Pontifical Catholic University, SP. She holds an MS in Law from the University of Oklahoma, Norman, USA. She was Visiting Researcher at the University of Oklahoma College of Law, Norman (2008–2009). Also, she was Visiting Researcher PRH04/ANP/MCTI/IEE/USP (2013–2016). Currently, she is Visiting Researcher PRH33.1/ANP/MCTI/IEE / USP, collaborating professor and advisor at PPGE/USP, and collaborating researcher at the Center for Gas Innovation (RCGI/FAPESP/Shell), and Deputy Director of CEPID/SHELL/FAPESP Energy Policy and Economics Program.

da Silva Vescovi Chedid Thaiz is a lawyer, researcher, advisor and professor; post-doctorate student in Energy from the Graduate Program in Energy of the University of São Paulo (PPGE/USP); has Ph.D. in Energy from the University of São Paulo USP in collaboration with Georgetown University Law School; Master in Fundamental Rights and Guarantees by the Vitoria Law School; Researcher at the Research Center for Gas Innovation, RCGI/FAPESP/Shell, at the University of São Paulo; publishing author of work focused on international and comparative energy law, Brazilian and UNCLOS petroleum royalties, climate change and carbon capture, Brazilian infrastructure and energy contracts as SLO, and other energy-related themes.

dos Santos Edmilson Moutinho is Associate Professor at the Institute for Energy and the Environment at São Paulo University (IEE/USP). He is also President of the Postgraduate Commission and Coordinator of the Economic and Energy Politics programme at the Research Center for Gas Innovation (RCGI) and also at São Paulo University.

Heredia Florencia is a partner of Allende & Brea Argentinean law firm leading the Energy & Natural Resources Department. Her practice focuses on mining law and environmental matters with 30 years extensive experience advising financing institutions and companies in complex mining and energy transactions in Argentina. She is an active member and having served in different positions in the RMMLF and the IBA. She has been repeatedly named as a leading professional in the areas of Natural Resources by Chambers & Partners, Who's Who Legal and Legal 500 and frequently named "Mining Lawyer of the Year".

Jaramillo Claudia Fonseca is a Partner at the law firm Garrido Fonseca Abogados. She focuses on advising and assisting clients in commercial, contractual and environmental law, as well as in arbitration and national and international negotiation. With more than 18 years of experience in legal practice, she is an expert in advising domestic and foreign clients in the areas of infrastructure, oil and gas, energy, and natural resources, having advised more than 50 companies in strategic issues of their business and operations. She has also participated in and supported more than 100 arbitration cases related to the energy sector. She has a law degree from Universidad de los Andes in Bogotá, Colombia (2002), and postgraduate studies in International Business with an emphasis on Capital Markets and Investment Banking from the School of Government, Finance and International Relations, Universidad Externado de Colombia, in a joint program with the School of International and Public Affairs (SIPA), Columbia University, New York (2004).

Lanardonne Tomás is a transactional and regulatory lawyer specialised in oil and gas, conventional and renewable energy, mining, and natural resources-related litigation and arbitration with more than 15 years of experience. He is a founding partner of MHR Abogados, a top tier law firm in energy matters according to Chambers and Legal 500. He obtained his law degree at the Universidad de Buenos Aires (Honors Diploma) and his LL.M. in Energy Law and Policy at the University of Dundee (With Distinction), as a Chevening Scholar. He co-founded the first Argentine Energy Law Journal (RADEHM) in 2014.

Martinez Agostina L. is an Associate in Allende & Brea Argentinean law firm in the Energy & Natural Resources Department. She obtained her law degree from Universidad Católica de Cuyo and holds a Master in Laws (LL.M.) with Distinction in Mineral Law and Policy from the University of Dundee (CEPMLP). She attended several postgraduate courses including the "International Oil & Gas Law, Contracts and Negotiations Part I and II" course organised by the Rocky Mountain Mineral Law Foundation (RMMLF) and the Association of International Petroleum Negotiators (AIPN) in Houston. She is an active member of the RMMLF and the IBA.

Mata-García César R. holds a doctorate (Ph.D.) in Law and a LL.M. in Petroleum Law and Policy from the University of Dundee (Scotland, UK) and a LL.B. from the Andres Bello Catholic University (Caracas, Venezuela). He has more than 20 years of experience in the energy sector and more than 15 years of experience in the academic field, including academic research. He has been professor and visiting scholar in various universities in Venezuela and abroad. Currently, he works as legal consultant, professor, facilitator, instructor and writer on energy legal topics at the national and international levels.

Mazzochi Juan Cruz is a transactional and regulatory lawyer specialised in energy, currently working as an Associate in MHR Abogados. He obtained a *summa cum laude* law degree and completed a Hydrocarbons Postgraduate Diploma, both in Austral University. Also, he currently lectures as an Assistant Professor in Environmental Law, Constitutional Law and Common Law at Austral University.

Meneses Carolina focuses her practice on the regulation of natural and extractive resources, with emphasis on mining and environmental law and administrative law. A graduate lawyer from Universidad Peruana de Ciencias Aplicadas (UPC), Lima, Peru, she obtained her Master's Degree in Law (LL.M.) with specialisation in International Mining Law and Policy from the Centre for Energy, Petroleum, and Mineral Law and Policy (CEPMLP), University of Dundee, Scotland, having earned her degree with distinction and awarded the Rio Tinto Scholarship granted by Rio Tinto PLC and the CEPMLP. She is also Assistant Professor of the Mining, Environmental, Hydrocarbons, Electricity and Gas Law courses at UPC. She has also participated in the International Training Course 'Economic Tools for Conservation' of the Conservation Strategy Fund at Stanford University, California, USA. Further, she has worked at Sumitomo and Nyrstar, the Agency of Environmental Evaluation and Control (Organismo de Evaluación y Fiscalización Ambiental—OEFA), was a Senior Associate at Rebaza Alcazar & De Las Casas, a member of the Environmental Area of Muñiz, Olaya, Meléndez, Castro, Ono and Herrera, as well as assistant of the Mining Area of CMS-Grau.

Moreno Luis Ferney has a Ph.D. in Law from the University of Louvain (Belgium), Master in Economic Law from the University of Louvain (Belgium), Specialist in Taxation from the University of the Andes (Colombia), Certificate in Public Service Regulation from the University of Florida (USA) and Lawyer from University Externado of Colombia. Currently, he is the Director of the Department of Energy Mining Law of the Externado de Colombia University, member of the Ibero-American Association of Regulation (Asier), member of the Ibero-American Administrative Law Forum (IAALF) and member of the Ibero-American Association of Energy Rights (ASIDE).

Nishizaki Ernesto Beltrán has a degree in Law from the "Universidad Panamericana", with a specialty in Administrative Law Institutions from the same university and a LL.M. in Energy Law & Policy from the Center for Energy, Petroleum and Mineral Law & Policy of the University of Dundee, United Kingdom. He is an Associate in Grupo de Asesores Legales y Técnicos de la Construcción (GALTEC) in Mexico City. His research interests focus on energy, natural resources and energy transition.

Pereira Eduardo Guedes is a professor of natural resources and energy law, a full-time scholar at the Siberian Federal University and part-time, adjunct and/or visiting scholar in a number of leading academic institutions around the world (including the University of West Indies, University of São Paulo, Strathmore University, Agostinho Neto University, University of Aberdeen, Externado University of Colombia and among others). He has been active in the oil and gas industry for more than 15 years and is an international expert on joint operating agreements. His experience in this area—both academic and practical—is extensive. He has practical experience in over 40 jurisdictions covering America, Europe, Africa and Asia. He concluded his doctoral thesis on oil and gas joint ventures at the University of Aberdeen. He conducted postdoctoral research at Oxford Institute for Energy Studies (University of Oxford, UK), the Scandinavian Institute of Maritime Law (University of Oslo, Norway) and the Institute of Energy and Environment (University of São Paulo, Brazil). He is also a managing editor for the African Journal on Energy, Natural Resources and Environmental Law and an associate editor of OGEL. He is also the author and editor of several leading oil and gas textbooks. Further information about his profile and publications can be found at: www.eduardogpereira.com.

Urtubey Valentina Surraco is an Associate in Allende & Brea in the Energy & Natural Resources Department. She obtained her law degree magna cum laude from the Universidad Católica Argentina. She holds a postgraduate degree on the legal framework of renewable energy in Argentina from the Universidad Torcuato Di Tella and a course certificate on enterprises and sustainability from the University of Oxford. She is a member of the Rocky Mountain Mineral Law Foundation (RMMLF) and the International Bar Association (IBA), is certified by the International Compliance Association and is also member of the Argentine Ethics & Compliance Association.

Velarde Enrique is a qualified lawyer who gained his law qualifications at the Pontificia Universidad Católica del Perú, a LLM in Mineral Law and Policy from the Centre for Energy Petroleum and Mineral Law and Policy (CEPMLP), University of Dundee (UK), and an MSc in Environmental Policy and Regulation from the London School of Economics (LSE) (UK). He has experience in mining law, and environmental and natural resources regulation, as well as public policies on natural resources, environmental and university higher education. He is also proficient in English and Italian.

Zapata Sánchez Ana Cecilia is an Attorney at Law from Externado University, Colombia. She holds the degrees LL.M. in Energy Regulation and Specialization in Energy Mining Law, both from Externado University, Colombia.

Abbreviations

ACERA	Chilean Association of Renewable Energies and Storage
AMC	Argentine Mining Code
ANC	Autonomous City of Buenos Aires
ASEA	National Agency of Industrial Safety and Environmental Protection for the Hydrocarbons Sector (Mexico)
BCRP	Central Bank of Chile
BESS	Battery Energy Storage System
BTU	British Thermal Unit
CAMMESA	Administrator of the Power Grid in Argentina (*Compañía Administradora del Mercado Mayorista Eléctrico S.A.*)
CCUS	Carbon Capture Utilisation and Storage
CEN	National Electricity Coordinator (Chile)
CENAGAS	National Centre for Natural Gas Control (Mexico)
CNE	National Energy Commission (Chile)
CNH	National Hydrocarbons Commission (Mexico)
CNIH	National Hydrocarbon Information Centre (Mexico)
CO_2	Carbon Dioxide
COCHILCO	Chilean Copper Commission
COD	Commercial Operation Date
COES	Energy System Operator (Peru)

CONUEE	National Commission for Energy Efficiency Consumption (Mexico)
CRE	Energy Regulatory Commission (Mexico)
CSR	Corporate Social Responsibility
CVP	Venezuelan Petroleum Corporation
E&P	Exploration and Production
EDF	Environmental Defence Fund (Mexico)
EMI	Environmental Management Instrument (Peru)
ESG	Environmental, Social and Governance Perspective
EV	Electric Vehicle
GDP	Gross Domestic Product
GHG	Greenhouse Gas
GML	General Mining Law 1992 (Peru)
GTM-NDC	Working Group of the Peruvian Government in charge of formulating the Nationally Determined Contributions(Grupo de Trabajo Multisectorial de naturaleza temporal encargado de generar información técnica para orientar la implementación de las Contribuciones Nacionalmente Determinadas) (Peru)
IDB	Inter-American Development Bank
IEA	International Energy Agency
ILO 169	Indigenous and Tribal Peoples Convention 1989 (No.169) (Colombia)
IRENA	International Renewable Energy Agency
LGSE	General Law of Electric Services (Chile)
LNG	Liquefied Natural Gas
MEM	Wholesale Electricity Market (Argentina)
MINEM	Ministry of Energy and Mining (Argentina)
MINEM	Ministry of Energy and Mining (Peru)
MMBD	Million Barrels Per Day
MMSCFD	Million Standard Cubic Feet Per Day
MPD	Ministry of Productive Development (Argentina)
NCRE	Non-Conventional Renewable Energy
NDCs	National Determined Contributions
NOC	National Oil Company
NPCAD	National prior Consultation Authority Direction (Colombia)
O&G	Oil and Gas
OECD	Organisation of Economic Cooperation and Development

OEFA	Agency of Environmental Assessment and Enforcement (Peru)
OPEC	Organization of Petroleum Exporting Countries
OSINERGMIN	Energy Regulator of Peru (*Organismo Supervisor de la Inversión en Energía y Minería*) (Peru)
PEMEX	Petróleos Mexicanos
PEN	Local Currency (Peru)
PPA	Power Purchase Agreement
PPEE	National Energy Efficiency Program (Chile)
R&D	Research and Development
REE	Rare Earth Elements
REN21	Renewable Energy Policy Network for the 21st Century
RER	Renewable Energy Sources (elsewhere this is normally abbreviated to RES) (Peru)
SEC	Superintendence of Electricity and Fuels (Chile)
SEIN	National Interconnected Electric System (Sistema Eléctrico Interconectado Nacional
SENACE	Agency of Environmental Certification for Sustainable Investment (Peru)
SENER	Ministry of Energy (Mexico)
SLO	Social License to Operate
SUNAT	Custom and Tax Authority (Peru)
TPES	Total Primary Energy Supply
UPME	Mining Energy Planning Unit (Colombia)
URT	Special Administrative Unit for the Management of Restitution of Dispossessed Lands
USGS	United States Geological Survey
UTM	Universal Transverse Mercator

Note on Units

Power units	The power using or generating capacity of devices is measured in watts, or more usually kilowatts (kW) (1 kW = 1000 W).
Energy units	The kilowatt-hour (kWh) is the standard unit by which electricity is sold—1 kWh is the energy produced/consumed when a 1 kW rated generator/energy-consuming device runs for 1 h. A megawatt-hour (MWh) is 1000 kWh. Similarly, 1000 MWh = 1 GWh and so on.

List of Figures

List of Tables

Part I

Introduction

1

Latin America: Renewables at the Crossroads of Multiple Drivers

Geoffrey Wood

1.1 Introduction

Latin America is primed for rapid renewable energy growth. This is not just an expectation of hope given the need for the region and indeed the world to urgently address the red line of climate change and shift towards a zero carbon energy system in the next few decades.[1] It is based solidly on a number of factors. The vast and diverse geography of the region is endowed with substantial renewable energy resources and Latin America has one of the largest shares of renewables, due in large

[1] Geoffrey Wood and Keith Baker, *The Palgrave Handbook of Managing Fossil Fuels and Energy Transitions* (Palgrave Macmillan 2019).

G. Wood (✉)
School of Law, University of Stirling, Stirling, Scotland, UK
e-mail: geoffreywood73@gmail.com

G. Wood and J. F. Neira-Castro, *From Fossil Fuels to Low Carbon Energy Transition*, Energy, Climate and the Environment,
https://doi.org/10.1007/978-3-031-00299-1_1

part to the distinctive historical development of hydropower and recently biofuels.[2] However, this belies the recent diversification into other renewable sources including wind power, bioenergy for power, solar power and geothermal that in 2020 had a combined installed capacity of 280GW.[3] Representing a growth of more than 60% compared to ten years ago, around a quarter of the region's energy now comes from renewables, and this does not begin to consider the region's potential to tap into solar and wind resources which accounted for only 16% of total renewable generation in 2020.[4]

Latin America is also host to some of the world's most dynamic renewable energy markets. Concentrating on new (non-hydropower) deployment in the region, investment in renewable energy sources has grown 11-fold since 2014, almost double the global average rate. This means that Chile and Mexico have joined Brazil for the first time on the top 10 largest renewable energy markets list,[5] showcasing the rapid evolution of the regional energy mix in terms of diversification. Over the period 2010–2015, over US$80 billion worth of investment (excluding large-scale hydropower) in renewable energy occurred, with US$16.4 billion alone in 2015, representing 6% of global total investment in renewables.[6]

Moreover, many countries in the region have set ambitious renewable energy laws, policies, regulations and targets to address climate change.[7] At the same time, it is important not to omit the critical role played by fossil fuels and the extractives—two major sectors that contribute to climate damaging greenhouse gas emissions and drive

---

[2] International Renewable Energy Agency (IRENA), *Renewable Energy Market Analysis: Latin America* https://www.irena.org/publications/2016/Nov/Renewable-Energy-Market-Analysis-Latin-America.

[3] Bruna Alves, 'Renewable Energy in Latin America—Statistics and Facts' (Statistica, 5 July 2021) https://www.statista.com/topics/5318/renewable-energy-in-latin-america/.

[4] Izei Atxalandabaso, 'Renewable Energy in Latin America: 5 Renewable Energy Trends Emerging from South of Rio Grande' (Rated Power, 16 April 2021) https://ratedpower.com/blog/renewable-energy-latin-america/.

[5] Renewable Energy Policy Network for the 21st Century (REN21), *Renewables 2020: Global Status Report* https://www.ren21.net/gsr-2020/.

[6] IRENA (n 2).

[7] REN21 (n 5).

climate change—in national energy mixes, government revenue streams, national development and energy security of the region. It is also important to not ignore the linkages between these. The same can also be said about commodities, still one of the main regional exports. According to the Economic Commission for Latin America and the Caribbean (ECLAC), in 2018 commodities were worth US$456 million.[8] These patterns of economic development have been at the core of the development model in the region for the last few decades, and especially when the commodities boom price occurred between 2010 and 2015.

True, the per capita greenhouse gas emissions profile of someone in Latin America is only a third of the average citizen of Europe or the United States. At the same time, Latin America is one of the most vulnerable regions to the impact of climate change, with impacts on GDP and associated employment and so on. Due to the importance of hydropower as one of the main sources of power generation, the impacts of natural events such as El Niño and La Niña will become aggravated, causing alternating severe floods and droughts (already being felt in the region), affecting commodities like agriculture and placing national electricity systems in a vulnerable position.[9] Thus, climate change is a very real threat to Latin America, its citizens and its development, a somewhat paradoxical position given that the same drivers of climate change (hydrocarbons and extractives) have also largely driven development in the region.[10] The proportion of the region's greenhouse gas emissions is overall small in relative terms with other parts of the world (albeit growing), but this will not stop the region feeling the brunt of climate change due to the global nature of the problem. Therefore, Latin America

[8] Notably, both renewable energy and fossil fuel sources (petroleum and gas) exist as natural resources available in the region. See: Economic Commission for Latin America and the Caribbean (ECLAC), *Economic Commission for Latin America and the Caribbean: A Graphic View—August 2018* https://www.cepal.org/en/publications/43889-economics-climate-change-latin-america-and-caribbean-graphic-view.

[9] Ibid.

[10] Without precluding the development needs for many countries, it is worth noting that 'although fossil fuels have powered the industrialization of many nations and improved the lives of hundreds of millions of people, another century dominated by fossil fuels would be disastrous', see Geoffrey Wood, Fossil Fuels in a Carbon-Constrained World, in Geoffrey Wood and Keith Baker (eds.), *The Palgrave Handbook of Managing Fossil Fuels and Energy Transitions* (Palgrave Macmillan, 2019), 3.

is at a crossroads: Should it make the drastic shift from fossil fuels to renewable energy sources?

This book, with its unique focus on new regulatory trends in Latin America as the region wrestles with the shift from a high carbon to a low carbon world as part of the ongoing energy transition, seeks to capture the contrasting and at times conflicting trends in energy law, policy and regulation in the region. By looking at the 5 key themes of hydrocarbons, electricity (power), mining, social license to operate and arbitration/dispute resolution at both the regional level and through in-depth country case studies (Argentina, Brazil, Chile, Colombia, Mexico, Peru, Venezuela), this book also attempts to address the complex issues involved via a holistic account of energy law, policy and regulation in Latin America. This is particularly relevant. The region is well known for its lack of political and legal stability, a factor that has specific relevance to the energy sector given the dominant role that energy plays in the Latin American economies. This book also aims to contribute to the current lack of focus on Latin America as a regional example of the opportunities and challenges inherent in low carbon energy transition, thus moving forward the debate on a relatively unexplored yet important part of the world.

This chapter introduces Latin America in all its rich diversity before exploring the complex contextual drivers that have led the region to a crossroads. This chapter ends by outlining the structure of the book, including a summary of the chapters contained therein.

1.2 From Mexico to Tierra del Fuego: Latin America

Latin America is a region of extremes. It is home to the world's largest river (the Amazon) and the world's driest place (the Atacama Desert).[11] It contains one of the world's largest tropical rainforests (the Amazon), vast grasslands (the Pampas), vast river basins, countless rivers and wetlands,

[11] National Geographic, 'South America: Human Geography' (National Geographic, n.d) https://www.nationalgeographic.org/encyclopedia/south-america-human-geography/.

a vast coastline that borders three oceans (the Pacific, Atlantic, and Southern) and the Caribbean Sea and the Gulf of Mexico, whole hosts of islands (both tropical and temperate), and the Andes mountain range that forms the backbone of the region, running north to south towards the continent of Antarctica beyond the tip of Tierra del Fuego (the Land of Fire). Latin America is also one of the most biodiverse regions in the world: around 60% of global terrestrial life (and diverse freshwater and marine species) can be found within Latin America.[12] At the national level, the region is home to 3 out of the top 5 countries globally with the most amphibian, bird, fish, mammal, reptilian and plant life. Indeed, the Amazon region alone is home to 10% of the Earth's biodiversity.

But what do we mean by Latin America? Geographically, this is the portion of the Americas south of the Rio Grande (the 'Big River') comprising many countries and regions from North America, Central America, the Caribbean and South America (see Fig. 1.1).[13] Highlighting the diverse geography, peoples and cultures, the region consists of 20 countries and 14 dependent territories covering 7.4 million square miles, almost 13% of the Earth's land surface, and home to an estimated 625 million people.[14]

Latin America is also a region of extremes for other reasons: colonisation by Europeans from the 1500s onwards, the decimation and displacement of the indigenous peoples, the African slave trade resulting in the prominent African diaspora and the wars of independence in the late eighteenth and early nineteenth centuries that culminated in independence throughout the region from the colonial powers by 1898.[15] Latin America is also home to extreme poverty and wealth, both within and between the nations and dependencies that the region is composed

---

[12] Jessica Carey-Webb, Latin America's Biodiversity is Critical for Global Goals (NRDC, 22 May 2020) https://www.nrdc.org/experts/jessica-carey-webb/lets-protect-latin-americas-biodiversity.

[13] Latin America refers to those countries where Romance languages such as Spanish, French and Portuguese are predominantly spoken. As such, it differs from the more commonly used geographical terms such of North America, Central America and South America (although it contains parts of all including the Caribbean). Originating in 1856, the term is broader than other commonly used terms such as Hispanic America and Ibero-America.

[14] The World Bank, *World Development Indicators* http://wdi.worldbank.org/table/3.1.

[15] Center for International Education, *Focus on Latin America: Geography and Culture* https://www.berea.edu/cie/focus-latin-america/.

Fig. 1.1 Geographical and jurisdictional range of Latin America[16]

of. Energy issues are further compounded by the prevalent conflicts between the energy industry and indigenous communities that continue today.

Natural resources are also distributed unequally between nations, and it is this that highlights the challenges and opportunities for Latin America in its efforts to achieve energy transition. In addition to vast renewable energy resources, the region is also endowed with vast resources of fossil fuels. Fossil fuels remain dominant in the regional energy mix because of Latin America's role as a global oil and gas

[16] Wikipedia, 'Latin America' (Wikipedia, n.d). https://en.wikipedia.org/wiki/Latin_America.

producer. This is clearly portrayed in Fig. 1.2. Oil (46%), gas (23%) and coal (5%) combined account for 74% of total primary energy supply.[17]

In particular, Mexico, Brazil and Venezuela are the 10th, 11th and 12th largest oil producers in the world, accounting for nearly 75% of the region's oil production.[18] Indeed, Latin America boasts the largest crude oil reserves globally.[19] Gas use has increased partly due to substituting oil in the power sector, and now accounts for 23% of total regional primary energy supply. Since the 1970s, gas production has increased steadily, with the two biggest producers, Argentina and Venezuela, being joined after 2000 by Trinidad and Tobago, Bolivia, Brazil, Colombia and Peru.[20]

Turning to renewables, as mentioned previously, Latin America has impressive shares of hydropower and bioenergy (mainly for the transport and industrial sectors).[21] As Fig. 1.2 shows, all renewables account for 26% of total primary energy supply. With the increasing expansion of other renewable energy sources, the share of both hydropower and (non-power) bioenergy has decreased in relative terms. The current renewable energy success story in the region is that of bioenergy for power (Brazil) and onshore wind (Brazil, Mexico, Uruguay and Panama), with solar power (Chile, Mexico, Peru and Uruguay) growing significantly albeit from a small base.[22] This diversification also occurs at the sub-regional level (Central America, Andean States, Southern Cone, and Mexico and Brazil) (see Fig. 1.2). All show exponential growth for other renewables.

---

[17] This is approximately the same proportion (70%) as found for the EU in 2019, and not far off the global average. See: Europa, Energy statistics—an overview (Europa, May 2021) https://ec.europa.eu/eurostat/statistics-explained/index.php?title=Energy_statistics_-_an_overview.

[18] J. William Carpenter, 'The Biggest Oil Producers in Latin America' (Investopedia, 5 December 2019) https://www.investopedia.com/articles/investing/101315/biggest-oil-producers-latin-america.asp.

[19] Bruna Alves, 'Leading Countries in Proved Crude Oil Reserves in Latin America and the Caribbean in 2021' (Statistica, June 2021) https://www.statista.com/statistics/961596/latin-america-crude-oil-reserves-country/.

[20] International Association of Oil and Gas Producers (IOGP), *Oil & Gas Production in Central and South America: Investment Needed to Meet Rising Regional Demand* https://www.iogp.org/bookstore/product/global-energy-brief-latin-america/.

[21] IRENA (n 2).

[22] The countries in brackets represent the regional leaders in the respective renewable energy sources, but all nations are deploying or making efforts to develop and deploy various renewable energy sources.

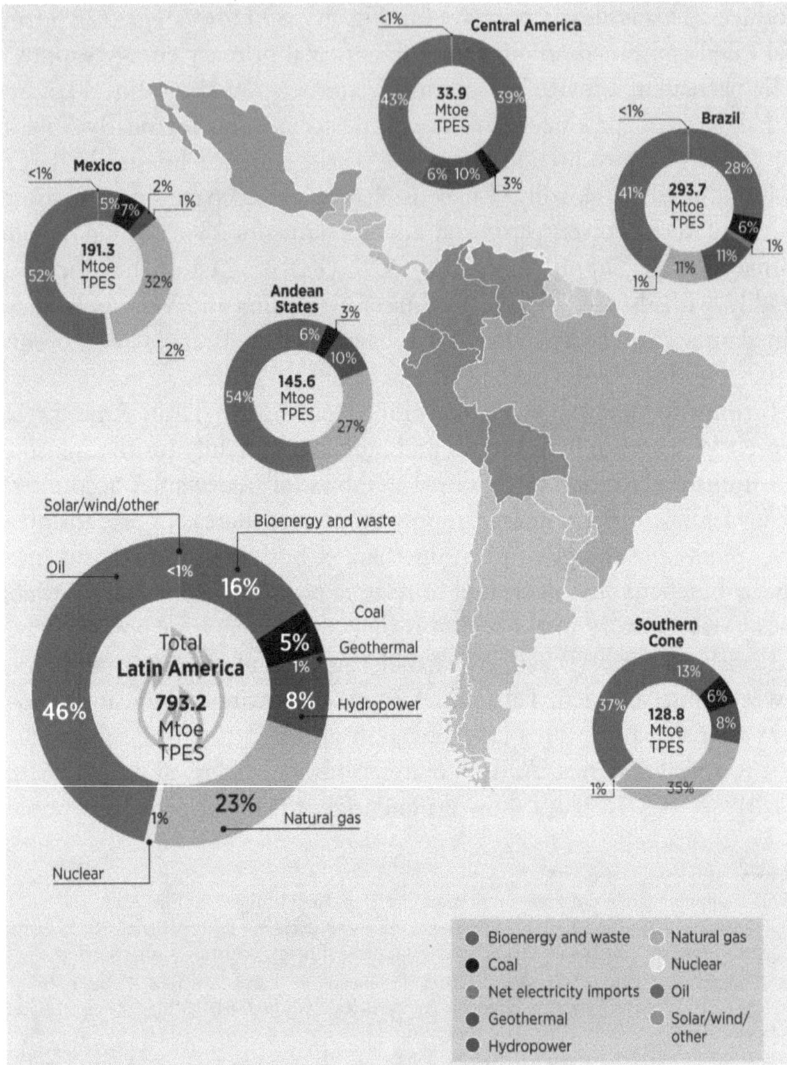

Fig. 1.2 Total primary energy supply by region (Latin America) and sub-region in 2013[23]

---

[23] IRENA (n 2).

This is important as such growth can benefit from the synergy between large hydropower and other renewables given that they are primarily intermittent in nature, thus requiring low carbon baseload generation to balance supply and demand and vice versa.[24]

Individual country success stories in Latin America should also be applauded. In 2019, Costa Rica generated an impressive 99.62% of the country's electricity from renewables; this was primarily from hydropower with wind power, geothermal energy and solar power making important contributions.[25] Other countries also show notable success in terms of the share of renewables in electricity generation: Brazil (83.3%), Honduras (60%), Peru (59.44%), Guatemala (59%), Nicaragua (50%) and Chile (22.8%).[26] Looking at total energy usage, Brazil sits second place in the global rankings.[27] Looking at renewable power across the region, installed capacity growth in non-hydropower renewables has more than tripled between 2005 and 2015, from 10 to 36GW. But it is important to focus beyond just renewable energies: with its extensive extractives potential, Latin America will likely play a critical role in supplying the minerals increasingly needed to build the infrastructure and provide the battery (storage) capacity required for energy transition, as evidenced by the Lithium Triangle states of Argentina, Bolivia and Chile.[28]

Diversification can also be seen by looking at regional investment trends in renewable energy. Prior to 2010, Brazil dominated Latin America as the primary destination for over 70% of the region's renewable energy investment.[29] By 2015, this share had dropped to a little

---

24 Ibid.

25 World Smart Cities Forum, '5 Places Running on 100% Renewable Energy' (World Smart Cities Forum, 2020). https://worldsmartcities.org/5-places-running-on-100-renewable-energy/.

26 REN21 (n 5).

27 Maeve Campbell, 'Which Country Is the World Leader in Renewable Energy in 2021' (Euronews, 16 August 2021). https://www.euronews.com/green/2021/08/02/which-country-is-the-world-leader-in-renewable-energy-in-2021.

28 Ryan C. Berg and T. Andrew Sady-Kennedy, 'South America's Lithium Triangle: Opportunities for the Biden Administration (Center for Strategic and International Studies (CSIS), 17 August 2021). https://www.csis.org/analysis/south-americas-lithium-triangle-opportunities-biden-administration.

29 IRENA (n 2).

over 40%. Importantly, this does not represent a reduction in renewable energy investment in the region per se. On the contrary, it highlights the success of other Latin American nations in attracting investment to deploy renewable capacity, notably Chile, Uruguay, Honduras, Peru and others. Indeed, regional investment in renewables has continued to grow as highlighted by the significant levels of investment in recent years.

A lot of the success is due to the decisive role of enabling policies, particularly those policy instruments that have helped reduce the costs of deploying renewable energy (with a particular emphasis on onshore wind and solar photovoltaic). Similar to elsewhere, Latin American policymakers are becoming increasingly aware of the socio-economic benefits of doing so additional to combatting global climate change, including GDP growth, the development of local industries and supply chains and the resultant jobs that come with this, alongside greater power system reliability.[30]

1.3 At the Crossroads: Moving Forward?

Latin America sits at a crossroads in the low carbon energy transition. The region is well-endowed in renewable energy resources and rich but heavily dependent on fossil fuels and extractives for energy security and economic development, both historical and current drivers of Latin American energy policy. On the one hand, growth in renewables is going from strength to strength, and the region is rich in renewable energy success stories. Energy mixes at the national and indeed regional level are diversifying away from a heavy reliance on the two Hs: hydrocarbons and hydropower. On the other hand, as this book shows, not all is rosy in the Latin American energy 'garden'. There are significant pressures on a Latin American energy transition. In addition to the pressures that derive from a long dependence on fossil fuels and extractives for

[30] Geoffrey Wood and Stephen Dow, 'What Lessons Have Been Learned in Reforming the Renewables Obligation? An Analysis of Internal and External Failures in UK Renewable Energy Policy' [2011] 39 *Energy Policy* 39(5) 2228.

political, economic and other reasons, renewables-driven energy diversification is still comparatively low, and not all countries in the region are equally embracing the need to address climate change and transition from a high carbon to a low carbon energy system. As Adnan Z. Amin, the first Director-General of IRENA, stated in the report 'Renewable Energy market Analysis: Latin America':

> In recent years, energy security has been a key driver for energy diversification to limit adverse macroeconomic effects due to the high reliance on fossil fuels and to reduce vulnerability to recurring climate events impacting hydropower generation. The imperative to decarbonise, together with national energy security concerns, in the context of rapidly falling costs of non-hydropower renewables, provides a compelling case for broader renewable energy development in Latin America.[31]

Besides energy security, then, other drivers are at play, including economic competitiveness and environmental sustainability. Put another way, renewables currently find themselves at a crossroads of drivers for energy diversification. This is the key question for Latin America: how to decarbonise in a way to address the drivers of energy security, economic competitiveness and environmental sustainability? Navigating this crossroad will require a delicate balancing act between these powerful drivers, not least to optimise the region's energy transition in terms of successfully incentivising and constructing the necessary renewable energy infrastructure, and designing the frameworks in which they will operate, but also to avoid consumers and the poorest/marginalised from bearing the brunt of the transition. As this chapter and subsequent chapters of the book hopefully show, Latin America is at the edge of a 'tipping point', and despite individual success stories the region as a whole still has some way to go in facing up to the realities of climate change and at the same time realising the social, economic and environmental opportunities that a shift to renewables can bring. Notably, the region is not alone in this predicament.

---

[31] IRENA (n 2) 3.

Caution, however, is required in generalising the findings of the book to Latin America as a whole, given the sheer diversity of the region in terms of geography, culture, politics and resources (as touched on above). However, it is hoped that this book will shine a light on the lessons that need to be learned to highlight the way forward for the region to move towards a low carbon energy future. Further, it is hoped that the insights provided in the book will provide valuable lessons for other regions across the world.

1.4 Outline of the Book

The book From Fossil Fuels to Low carbon Energy Transition: New Regulatory Trends in Latin America consists of four parts, of which Part I contains the introduction to the book. Part II looks at the themes of hydrocarbons, electricity (power) and mining whilst Part III looks at the concept of social license to operate and dispute resolution mechanisms in the region. Part IV provides an epilogue to the book.

In Part I, this chapter by Dr Geoffrey Wood sets out the context of Latin America as it shifts from fossil fuels to renewable energy sources, with a detailed overview of the region through the dual perspectives of energy transitions and a dependence on hydrocarbons and minerals. This chapter also explores the complex set of drivers underlying energy transition in the region. Chapter 2 by Luis Ferney Moreno and Daniela Aguilar Abaunza takes a high level overview of Latin America to explore the ways in which energy law has been transformed to become a sustainable, unified and global right, particularly in the context of Latin America. As such, this chapter fills a gap in the extant literature about what energy law involves through a reflection of its different stages as evidenced in Latin America.

Starting off Part II, in Chapter 3, Abel M. Venero Carrasco looks at the Peruvian renewable energy experiment and its lessons for policymakers. The introduction of specific regulations to incentivise the development of new power generation technologies is a sensible regulatory decision, but policymakers should take care not only to define a well-structured set of rules, but also to define clear objectives in order to assess the success of the policy. Additionally, policymakers should revise the impacts of the rules on the current power sector and verify if these impacts would be offset by the benefits. This chapter argues that in the case of the renewable energy regulation introduced in Peru in 2010, none of the above has been made. In particular, the costs associated with the regulation cannot be properly compared to the benefits sought and, more harmfully, the interaction of these rules with the regulatory framework in place has created unintended consequences mostly endured by the smallest of the power consumers. The chapter concludes that policymakers have failed to learn from past experiences and are trying to maintain a set of rules that clearly do not deliver sound policy.

Chapter 4 by Andrés Zamorano analyses Chile's search for a regulatory framework and long-term policies capable to bring security of energy supply to the country. Set within the context of an open economy with a liberalised electricity market, Chile has implemented a series of reforms and regulations in the field of energy policy capable of bringing long-term stability to reach foreign investment and strategic cooperation that would allow, in the long term, to provide through the implementation of technologies related to Non-Conventional Renewable Energy (NCRE), native and low-cost sources of fuel to achieve energy independence and security of energy supply. So far, efforts have been fruitful and the country has already managed to meet its self-imposed goal of generating 20% of its electricity with NCRE by 2025, whilst keeping in mind that these advances are only the first part of the road ahead.

Chapter 5 by César R. Mata-García turns to the regulation of the Venezuelan petroleum industry. This chapter argues that over the last two decades the Venezuelan hydrocarbons industry has been affected by multiple political, social, economic, technical, legal and even dogmatic factors which have compromised its performance and productiveness, including declining levels of crude oil production from 3,1 MMBD to just above 480 MBD. One of the Venezuelan petroleum industry's main legal problems is excessive control and regulation by the State, which has not only been used—in a dogmatic manner—as an instrument of political and social manipulation, but also as a tool to influence investments which are considered contrary to the interests of the 'ideological' plans of the government of the day. The author identifies and reflects on how excessive State control and regulation and the lack of professional and opportune decision-making in the Venezuelan hydrocarbons industry have affected the maintenance, the stabilisation and the development of said sector and, therefore, the legal certainty and stability requested by the investors. In concluding, the chapter warns that the petroleum sector should use the learned lessons from the past and apply them to avoid repeating the political, social, economic, technical, legal and dogmatic consequences.

Chapter 6 by Ernesto Beltrán Nishizaki focuses on Mexico, one of the largest oil producers in Latin American. This chapter argues that the importance and relevance for national development are crucial to understanding the Mexican ideology and ideocracy linked with the oil and gas sector. Even though oil and gas have been associated with progress, climate change and the current circumstance should create a new paradigm in which the centre of public policies is focused on the environment, sustainability and a better future for subsequent generations. If the Mexican Government consider them as the centre of public politics, according to international trends, the energy transitions shall play a key role. PEMEX as the National Oil Company (NOC) and the major player in the Mexican oil and gas industry must consider the energy transition as a part of its modernisation process. As the chapter unfortunately shows, PEMEX is not considering it as part of its strategic plans, despite other major oil companies doing so and governments and wider society claiming the need to do so with more impetus every day. This chapter aims to analyse, from a historical to a practical perspective, the role of the Mexican NOC in the energy transition and the challenges that the government may face if the energy transition is to progress in the following years.

Tomás Lanardonne and Juan Cruz Mazzochi explore the role that Argentina has to play in the ongoing energy transition. Chapter 7 argues that the country's gas, solar, wind, hydrogen and lithium resources can be exported to contribute to developing Argentina's net zero economy. This chapter studies the past, present and future of the Argentine energy mix and energy policy, with a focus on the opportunities and challenges that Argentina will face during the following decades.

In Chapter 8, Florencia Heredia, Agostina L. Martinez and Valentina Surraco Urtubey address the important role of the mining industry to the ongoing energy transition, considering that the said transition towards a low carbon economy is expected to be much more mineral intensive than previous transitions. From amongst a myriad of aspects that could be explored on this premise, this chapter focuses primarily on the phenomenon of the lithium sector in Argentina (one of the Lithium Triangle states). Certain minerals have acquired momentum under the energy transition and certainly lithium is one of them. This chapter aims to provide an overview of the current Argentine legal and regulatory framework, with the aim to discuss the opportunities and challenges ahead to develop lithium projects in Argentina.

In the final chapter of Part II, Enrique Velarde critically investigates whether the regulatory approach for energy transition in the Peruvian mining sector will adopt a regulatory approach or be based on private self-regulation. As Chapter 9 argues, nowadays the world recognises the energy and mining sectors as essential activities for the economy and providers of some not inconsequential benefits such as potential economic growth, employment, local income and development, and private investment. In that regard, Peru is a country of ancient mining tradition, which has been preserved and cultivated by leading international companies that have been expanding the country's mining activities. At the global and Latin American level, Peru is amongst the main producers of various metals (gold, silver, copper, lead, zinc, iron, tin, molybdenum and tellurium, amongst others). This scenario reflects not only the abundance of resources and production capacity of the Peruvian mining industry, but also the stability of Peru's economic policies, turning Peru into one of the key destinations for foreign investors on mining projects. However, the energy sector is simultaneously going through a transition stage to more renewable energies, and such energy transition is only achievable via strategic, coherent legislation that facilitates responsible investment and the exploitation of resources. Consequently, this chapter's focus is to discuss energy policy with emphasis on new regulatory developments in Peru, to give the reader a broad panorama of how the mining sector and the energy transition are regulated in the country.

Part III commences with Chapter 10. By carrying out an analysis of social license to operate and social justice in energy transition according to Brazilian legislation and practices, Thaiz da Silva Vescovi Chedid, Eduardo Pereira, Edmilson Moutinho dos Santos and Hirdan Katarina de Medeiros Costa set out to describe how the concept of social justice, consultation and benefits for the community addresses the needs of the people to allow the energy transition and the concerns that have generated interest in social and environmental public policies inserted into the national oil and gas sector. Using the deductive method and an exploratory research approach, this chapter seeks to introduce the reader to the topic with a focus on the Brazilian energy sector as the contextual background. The focus of the first part of the work, namely the social licensing to operate shows the social-environmental context and the challenges involved. Therefore, this chapter takes care to present the importance of social justice in the Brazilian energy sector, including socio-environmental protection, mitigation measures and public consultation in accordance with Brazilian law. In the development of the case, possible scenarios are brought to light alongside the lessons to be learned by Brazil.

In Chapter 11, Luis Bustos, Ana Cecilia Zapata Sanchez and Luis Fernando Bastidas Reyes carry out a review of the theory of social license to operate using Colombia as the case study. The chapter starts by reflecting on the theory, its characteristics and definitions, before subsequently investigating the relation between the SLO and public policy issues with respect to the mining and energy sector in Colombia. In particular, the chapter looks closely at three issues closely related to social license to operate and its use by companies in the country: (i) the participation of the territorial local authorities in the decisions of the energy sector; (ii) land restitution public policy; and (iii) the application of Prior Consultation. This approach is adopted to cast light on the use that has been given to this theory in the Colombian framework, and in doing so identifies some challenges that will have to be overcome in the coming years.

With a focus on arbitration and dispute resolution in Peru during the energy transition, Carolina Meneses, Luis Miguel Elías and Alejandra Galvez argue in Chapter 12 that the energy transition presents a series of legal and business risks that will probably be addressed through arbitration. It is essential to understand the development of arbitration clauses, arbitrable risks and precedents, and the identification of potential cases that can come up in the development of the energy transition in the region. This chapter looks at these issues using renewable energy resource regulation in Peru through an analysis of the regime based on arbitration awards in order to seek ways to optimise the deployment of renewables and thus the energy transition not just for the companies involved and the State, but also for citizens who will ultimately pay for it.

Part III concludes by looking at how the energy transition must be organised. In Chapter 13, Claudia Fonseca shows that by virtue of the environmental commitments in terms of reducing greenhouse gas emissions throughout the world, there is a race against time to achieve the desired energy transition to produce and utilise cleaner energies. Claudia argues that this 'race' must be organised, with a clear schedule both for each state and for each investor who intends to invest in the energy transition, so that, throughout the lifecycle and performance of long-term energy projects, no surprises will arise that make investors feel disappointed in their economic expectations. Further, states must not make decisions to meet the commitments to reduce emissions for which they may be subsequently sued.

In Part IV, Ignacio Herrera Anchustegui provides an epilogue to the book, drawing on the overarching findings and conclusions from the book chapters in Chapter 14. In particular, Ignacio reflects on energy law and energy governance in Latin America through a discussion of three issues: resource abundance; renewable energy investment; and renewable-driven energy justice. This chapter concludes that the energy transition in the region must be clean and rapid, but more importantly, inclusive and fair.

2

Transformation of Energy Law in Latin America: Towards a Sustainable, Unified and Global Energy Law

Luis Ferney Moreno and Daniela Aguilar Abaunza

2.1 Introduction

Energy law in Latin America is undergoing a transformation from a dispersed and autarchic law focused on the organisation of the industry to law in pursuit of the sustainability and unification of global energy. This transformation makes this discipline transcendent at a positive law level to such an extent that it must be recognised that it has grown to a new dimension. In the context of Latin America and even Ibero-America, the offer of countless academic programmes that are training

L. F. Moreno (✉) · D. A. Abaunza
Department of Energy Mining Law, Externado University, Bogotá, Colombia
e-mail: luis.moreno@uexternado.edu.co

D. A. Abaunza
e-mail: daniela.aguilar@uexternado.edu.co

G. Wood and J. F. Neira-Castro, *From Fossil Fuels to Low Carbon Energy Transition*, Energy, Climate and the Environment,
https://doi.org/10.1007/978-3-031-00299-1_2

professionals and at the same time are dedicating to research on issues that have to do with this specialty.

We propose to write this article to fill a gap in the doctrine about what energy law constitutes. Above all, we want to fill a gap about which transformations have this discipline had. Energy Law is a branch that has gained popularity, but it has not been written in a systematic, doctrinal and dogmatic way that truly makes up this discipline. For example, mining law, oil law, gas law and electricity regulation are discussed in greater depth, but not systematically on the concept of energy law. Precisely, the purpose of this article is to raise awareness about integrating this concept from its original conception, of a dispersed and autarchic law, to its recent conceptualisation as a unified and global sustainable energy law, particularly in the context of Latin America.

2.2 Disperse and Autarchic Energy Law: Pre-2010

Before 2012, Energy law was a dispersed law. That affirmation is because it was shaped with separate disciplines such as mining law for coal, oil law, gas law and electricity law. It must be recognised that there was not a synergy or principle that unified them. Those disciplines all came together around the world's energy law, but each followed its own path and always sought its independence.[1] The previous statement can be seen in legal practice, where oil law specialists only focused on that discipline and had no interest in the others. Nevertheless, on the other side, there were specialists in electricity law or regulation not understanding each other. A certain elite bonus was generated regarding oil law since it's humanity's most consumed energy source to this day.

Another element that characterises the energy law before the year 2012 was the marked tendency to be autarchic. That is, to concentrate its interest in determining the regulations only for the organisation of

[1] González Márquez, J. J. (2017). *Nuevo derecho energético mexicano.* Ciudad de México: Universidad Autonoma Metropolitana.

the coal, oil, electricity and gas industries without taking other considerations into account. However, it must be recognised that with that strenuous self-sufficiency without looking at other aspects, the regulation contributes to the development of these sectors. Indeed, inside each discipline, it has formed its institutions or governance, sources, subjects and objective about the regulations and essential legal institutions. Another thing to mention is that before the 90s, the regulation was based on an entrepreneurial state, in which the state directly interfered with the industry. Long after, in the 90s some countries freed and privatised their industries, the reason why the state's role changed to become a regulatory state in this sector.[2]

Within this context, this discipline of law that is interrelated with public law and private law, was born for Latin America, initially from legal practice. However, since the 90s, it has been incorporated into the academic world. The energy law does not originate from the systematic construction of the doctrine. It has more likely been a product of the legal praxis of lawyers. Later, since the 90s in Ibero-America, the postgraduate programmes began offering energy law-related courses in certain universities in Latin America. At the time that the postgraduate programme of specialisation in Energy and Mining Law at the Externado University in 1998 was established (which began in 1999), there was only natural resource law programme in Argentina at the postgraduate level. In Spain, at that time, there was no postgraduate programme on this subject. Today, there are several postgraduate programmes in energy law in Latin America and Ibero-America. This initial concept of what energy law is has changed since 2010, as we can see in the next point of this article, which in effect leads us to a new conception of the energy right.

---

[2] Ariño Ortiz, G. (1999). *Principios de Derecho Publico Economico.* Granada: Comares: Editorial SL.

2.3 Sustainable, Unified and Global Energy Law: Post-2010

Since 2010, energy law has been transformed: first, it has become a sustainable energy law due to the strong introduction of the principles of the sustainable development goals and the measures against climate change. Secondly, it becomes a more unified energy law, due to the energy transition, and thirdly, it becomes a global energy law due to the internationalisation of regulation.

a. Sustainable energy law

Regarding the configuration of a sustainable energy law, as already pointed out, this is due to several important events such as: First, the World Energy Council postulates the energy trilemma in 2010, which included: energy security, environmental sustainability and social sustainability. Secondly, the proclamation by the United Nations General Assembly of 2012 as the International Year of Sustainable Energy for All. Third, but not least important, it is the declaration of the United Nations Sustainable Development Conference held in Rio de Janeiro in 2012. The final agreement title was "*the future we want*",[3] which, for instance, was adhered to by many Latin American countries on their way to find a green economy and sustainable development goals. For example, Colombia adheres willing to reach sustainable development such as access to energy and clean energy. Later, incorporate it through the Conpes Document 3918 of March 2018, adopting strategies to implement these objectives.

Additionally, Mexico being part of the countries that firstly signed the declaration was also awarded as the International Environmental Leader for its outstanding work in the approval of the General Law on Climate Change. Moreover, Uruguay published the Regional Agreement on Access to Information, Public Participation and Access to Justice in Environmental Matters in Latin America and the Caribbean, approved

[3] United Nations Sustainable Development Conference Rio de Janeiro. (2012). *Final Agreement: The Future We Want.* Rio de Janeiro.

by Law No. 19,773 of 2019 in article 1 after Escazu. It should be clarified that the document originated from the United Nations Conference on Sustainable Development (Rio + 20).

It is worth highlighting that the 2015 United Nations conference on climate change COP 21 held in Paris, in which commitments were established between countries with the aim of reducing global warming, contributing to efficient energy use and helping the generation change from fossil to clean sources. Latin America, Colombia, Bolivia, Chile and Argentina (among others) developed tools and institutional aspects that together combine with the purpose of enacting the aforementioned commitments. Colombia started with the Law 1715 of 2014 and Law 1931 of 2018, both for climate change mitigation. Bolivia expedited Law 604 from 2014 which regulates risk management, measures that include risk reduction through prevention, mitigation and recovery of the environment. Additionally, Bolivia also arranged through Law 1182 of 2019 justice for environmental matters, ratifying Escazu's agreement.

Furthermore, Chile expedited a regulation for the Registry of Emissions and Transfers of Pollutants in 2013 and a new Energy Policy in 2015. Both aiming to change their fossil sources of generation and cleaning their industry. Finally, Argentina has also been introducing policy in an effort to commit to its climate change goals. These include, introducing Regulation of the Law of Minimum Budgets for Adaptation and Mitigation to Climate Change by Law No. 27,520 in 2020, and the Law of Minimum Budgets for Adaptation and Mitigation to Global Climate Change included by Law No. 27,520 of 2019. All the above-mentioned policy move us from simple, plain energy legislation to a more sustainable energy law.

b. Unified Energy Law

Regarding the transmutation of energy law into a unified energy law, we consider that this occurs due to the clean energy transition context.[4]

---

[4] Heffron, R. J., Rønne, A., & Tomain, J. P. (2018). A Treatise for Energy Law. *Journal of World Energy Law and Business*. Oxford University Press.

Since 2010, this has been universalised with the proposal of the decarbonisation of energy through progressive abandonment of fossil fuels (coal, oil and natural gas) and the electrification of energy to a greater extent. Such a change with regard to the greater use of electricity puts us in need of diversifying sources. The competition of energy sources with a preponderance of electricity paradoxically leads us to a true unification of energy law. Such an affirmation is based on today's reality. It is impossible to look in an isolated manner at the regulation of each of the primary and secondary energy sources. This is because they can be substitutes within these new objectives and energy transition context. That is why a convergent regulation of energy sources is required. Energy transition and climate change justifies the transmutation of energy law into a unified energy law (oil law, gas law, electricity law and unified and convergent renewable energy law).

We emphasise that decarbonisation is the starting and finishing point of the energy transition, which begins in Latin America with energy efficiency, meaning the rational and efficient use of energy consumption. Indeed, after the electricity and hydrocarbon bills for liberalisation and privatisation promoted by the World Bank in the 90s, laws relating to the efficient and rational use of energy gained approval around the period 1996 to 2001, promoted by the Economic Commission for Latin America and the Caribbean.[5] In the case of Colombia, the Law 691 of 2001 promotes rational and efficient use of energy by creating programmes that encourage energy efficiency measures throughout different economic sectors.[6] In Brazil, Law 10.295 of 2001 began mandating the national government to incorporate a national policy for the conservation and rational use of energy. Another example is Costa Rica, with the enactment of the Law 7447 of 1994, the first law in the Latin American region promoting energy efficiency measures. However, energy efficiency policy was not enough for the decarbonisation of the sector because clearly, energy consumption was increasing.

[5] Moreno Castillo, L. F. (2019). *Energia renovable y eficiencia energetica en Americla Latina: una vison regulatoria. En la proyección del derecho administrativo Peruano.* Perú: Palestra Editores.

[6] In Colombia, Law 697 of 2001 (Law of rational and efficient energy-URE) is approved.

Later, production of non-conventional renewable energy sources began to be promoted in Latin America, and therefore, various tax incentive laws were passed from 2001 to 2007. Since 2007, special laws on renewable energy have been approved with a general focus on electricity, particularly solar and wind power and for hydrocarbons.[7]

For instance, in Brazil Law 10.438 of 2002, known as Incentive Program for an Alternative Source of Energy (PROINFA) was enacted to increase the participation of wind energy, biomass and small hydro plants. In Peru, since the enactment of law 28832 of 2006, feed-in tariffs and special auctions began for renewable energy generation. In Colombia, Law 1715 of 2014 established various tax incentives for non-conventional renewable energy projects developers. At the same time, replacing fossil fuels with renewable sources was developed through so-called biofuels (fuel alcohol and biodiesel) and, the enactment of law and regulation aimed at promoting the production and demand of such fuels. Colombia enacted Law 693 of 2002 which created incentives for the production, trading and use of biofuels. In Brazil, Law 11.097 of 2005 established a blending percentage of biodiesel with diesel and later the Law 11.116 of 2005, established different tax incentives for the producers of biofuels.

One last stage is the proposal of a progressive abandonment of fossil fuels (coal, oil and natural gas) used for the greater part of electricity generation and for the promotion of other clean technologies beyond renewable energy. That is the case in Colombia, where recently Law 2099 of 2021 mandates the government to create incentives designed to promote and regulate hydrogen and carbon capture storage technologies. It is also worth noting that the expected electrification comes with five electricity sector transformations, which will become the most important secondary energy source in the coming years. These five transformations of electricity are: decarbonisation of electricity through renewable energy, the introduction of digital transformation, storage, energy efficiency and decentralisation of electricity.

[7] The author proposes that renewable energy and energy efficiency are predicated both for the electricity market and the hydrocarbon market, both on the supply and demand sides.

Increasing participation in renewable sources for electricity generation is needed because, although electricity derived from the use of renewables produces lower CO_2 emissions than fossil fuels, this mode of electricity generation can emit quite a bit depending on the source used. That is the importance of using clean sources for the generation of electricity. Energy storage is the process of maintaining energy in a form that permits the power to be made available in a useful form later in time, which was not even thinkable at the beginning of the electricity sector. It was always possible to store the source but not the energy once obtained. However, this possibility is now a reality, mainly on a large scale thanks to electric vehicles and lithium-ion batteries. Furthermore, energy storage is vital for the rise of renewable energy in the electricity mix, and correspondingly, there is a need for flexibility and balance in the grid management.[8] Regarding digitalisation, it refers to the series of digital technologies that allow the two-way flow of communication and electricity among the participants within the electricity system in an interactive and coordinated way, matching the energy needs and capabilities of the electricity system in the most efficient way.[9] Energy efficiency refers to measures to increase the amount of service from every unit of energy. Currently, it is the primary mechanism for reducing CO_2 emissions in the energy sector, beyond the share of renewables or demand-side management, and it is the most cost-effective means available for reducing emissions in both the short and long term.[10] Finally, decentralisation means that the power is generated at the consumption point or near it (as opposed to a traditional-centralised electricity system where the generators are located far away from the consumption points). Decentralisation means that instead of consumers passively receiving energy from the grid, they can now inject energy into the system on the distribution side.[11] Therefore, this reduces the cost and dependency of transportation (mainly on

---

[8] IEA. (2017). *Tracking Clean Energy Progress: Energy Technology Perspectives 2017: Informing Energy Sector Transformations.*

[9] IEA. (2011). *Technology Roadmap: Smart Grids, I.* International Energy Agency, Paris, 6.

[10] IEA. (2014). *Capturing the Multiple Benefits of Energy Efficiency.* International Energy Agency, Paris, 55.

[11] Salama, M., & El Khattam, W. (2004). Distributed Generation Technologies, Definitions and Benefits. In *Electric Power Systems Research* (pp. 119–120).

the transmission grid) and participates actively in the energy market by providing services to the grid.[12]

The above paradoxically places us within a unified energy law scenario. This means a convergent grouping of oil, gas, electricity and renewable energy laws, together with even a mining law for the integration of minerals used in the energy transition. It is necessary to consider that this unification, integration and convergence is consolidated from the configuration of shared principles, a shared regulatory cycle and likewise, a common interdisciplinary and normative source, and a convergent energy governance.[13] These factors are described further below.

2.3.1 Shared Principles

The main shared principles identified in energy law are: the principle of energy security, the principle of sovereignty of energy resources, the principle of sustainable development, the principle of mitigation of climate change, the principle of essentiality of energy, the principle of law of private and public property, principle of market foundations, principles of open access to infrastructure.[14] Guayo Castella proposes other principles such as the principle of resilience, the principle of economic efficiency, the principle of accessibility and affordability.[15]

2.3.2 Shared Regulatory Cycle

The shared regulatory cycle of the unified energy law begins with the regulation of the natural resources used as a primary energy source. This process then goes through the regulation of the infrastructure used to connect the points of production with the points of consumption. Finally, it ends with regulating public service activities so that these can

---

[12] Aguilar Abaunza, Daniela. (2020). *The Law for the Emerging Active Role of Electricity Consumers* (Thesis, Doctor of Philosophy (PhD). Hamilton: The University of Waikato.

[13] Parente, A. (2010). *Principios de derecho europeo de la energía.* Pamplona: Aranzadi.

[14] Heffron et al. (n 4).

[15] Guayo Castella, I. (2020). Concepto, contenido y principios del derecho de la energía. *Revista de Administración Pública*, 212, 309–346.

be provided regularly and continuously. Indeed, in its first cycle, natural resources such as minerals, oil, gas and water are regulated as such. It should be clarified that electricity is not a natural resource, but natural resources are used for its generation. The line of regulation for a natural resource is directed at an optimal use of these resources, their conservation and prevention of environmental deterioration. The second cycle, the regulation of infrastructure, involves, in a strict sense, inputting the points of production in contact with the points of consumption. In mining, oil, gas and electricity, an infrastructure is put in place that enables the supply of services and assets. The regulatory outline of these infrastructures is fundamentally based on achieving their best quality and, the right of third parties to access these infrastructures or facilities. Finally, the third cycle concerns public services.

There are activities in the oil, gas and electricity sector that are declared as public services. In the case of Colombia, the transport and supply of oil is considered a public service (Legislative Decree 1086, 1953); while for the electricity sector, according to the Law 142 and 143 of 1994, all the activities involved are considered public service (generation, transmission, distribution and retail). In the gas industry, all the activities have this category except for production (Law 142 of 1994). Considering this, we could analyse other countries in Latin America in which they either have the same vision as Colombia or take a different view on it. In Peru, the organic hydrocarbon Law (LOH, for its Spanish acronym) No. 26221 and amendments contemplate the exploitation and exploring of hydrocarbons as a statal matter, meaning others could interfere in it through concessions.

Nevertheless, Peru considers distribution and commerce as a public service. The mentioned policy highlights the general rules for all hydrocarbon activities in the country and establishes that the state will promote free competition and free access to economic activity. This is true so long as it accomplishes the purpose of achieving human well-being and national development (Organic hydrocarbon law No.26221). Along with the previous laws mentioned above, it is important to mention Law No. 23406 of 1992, which regulates everything relating to electricity as a public service. However, in the case of Chile the General Electric services Law, Law D.F.L. N° 1 from 1982, defines the supply as a public service. Furthermore, under Brazilian legislation according to Law Nº 9.427 of 1996, the energy sector is a public service whose regulation and concession regime is on the National Electric Power Agency Head. Brazil, considering Law Nº 9.478 of 1997, defines gas distribution by pipeline and local services for the sale of petroleum as a public service, in which private parties can participate.

The regulatory profile of the activities declared as public services, then, focuses mainly on those that must be provided regularly and continuously. However, this declaration implies that in some countries this activity is reserved to the state (state ownership).

2.3.3 Interdisciplinary

All the unified energy laws (oil, gas, electricity and renewable energy laws and even the mining law for minerals used in the energy transition) are interdisciplinary because they are areas where many public laws converge. On one hand, regulation, public contracts and state supervision, and on the other, private law or contracts between companies. Likewise, mining engineering, petroleum engineering, electrical engineering and even economy converge.

Moreover, regarding the legislative process, a monistic conception has prevailed in modern law, where exclusivity is given to legal regulations issued by the state. As affirmed by Jacques Chevalier, this conception is surpassed by the presence of multiple actors and norms.[16] Hence we find, in addition to state norms, supra-state (supranational) norms of international public organisations. These are extra-state norms of private international organisations (for example, the international standards of the oil and mining industry) and infra-state norms, originated by national private organisations through self-regulation.[17,18] There is also a recognition that state norms are norms of hard law such as the

[16] Chevallier, J. (2011). *El Estado posmoderno.* Bogotá D.C: U. Externado de Colombia.

[17] Black, J. (2001). Current Legal Problems. In *Decentring Regulation: Understanding the Role of Regulation and Self-Regulation in a Post Regulatory World.*

[18] Barton, B., Alastair, L., Barrera-Hernandez, L., & Ronne, A. (2006). Regulating Energy and Natural Resources. In B. Barton, *The Theoretical Context of Regulation* (pp. 12–14). Oxford: Oxford University Press.

law, decrees, resolutions or circulars, as well as norms of soft law such as directives, guidelines or recommendations.[19] An example of the latter being the different guidelines that have been adopted to promote energy efficiency in Colombia.

2.3.4 Convergent Energy Governance

Given this configuration of unified energy law, a model of unified energy governance is required that looks at the different sectors in a convergent way. This is because there is evidence of market convergence, which makes a unitary energy governance necessary. Such energy governance means establishing convergent regulation in the energy sector, for example, as seen in the telecommunications sector, a process known as sector coupling.[20] In the telecommunication sector, the sector coupling is due to technological convergence. This is different to the energy sector; in that it would be due to market convergence.

2.3.5 Global Energy Law

Like many other branches of law, energy law is also immersed in what is called today "global law". As a product of globalisation, it is impossible to ignore the multiple exchanges of regulations within the energy sector. This is especially true in the exercise of comparative law and the transposition of norms for the internationalisation of companies that operate in this energy sector around the world. However, as pointed out by Garcia-Manzano Jiménez, the international subjects that approve norms have been multiplied: either of hard law or soft law.[21] Such multiplication is not only due to an increase in international public organisations but also because of the proliferation of private international organisations.

[19] Black, J. (2002). "Critical Reflections on Regulation". *Australian Journal of Legal Philosophy*, 1–2.

[20] Fleming, R. (2020). "Clean or Renewable—Hydrogen and Power-to-Gas in EU Energy Law". *Journal of Energy & Natural Resources Law*, 16–28.

[21] Garcia-Manzano Jiménez. (2013). *Los reglamentos de las administraciones independientes*. Madrid: Editorial Civitas.

The proliferation of an international regulatory system that sustains global energy law is a reality that is to such an extent we can affirm it. There are even examples of specific public international energy-related organisations that could demonstrate that affirmation with their behaviour. These energy-related organisations are constantly issuing norms that could impact vertically, and they are either hard law or soft law rules.[22]

This depends on the capacities assigned to the international instruments of these international public organisations, such as the UN framework conventions recommendations of the International Energy Agency (IEA). For instance, in Colombia the Law 23 of 1988 ratified a treaty between Colombia and Canada for the peaceful cooperation in uses of nuclear energy in the context of a recommendation made by the International Atomic Energy Agency (IAEA).

Furthermore, recommendations made by the International Renewable Energy Agency (IRENA) have been implemented in domestic law. In the case of Colombia, Law 1665 of 2013 approved the Statute of IRENA. It is also the case of Chile with Law 21.305 of 2021, in which the regional action plan proposed by IRENA, designed to accelerate the deployment of renewable energy in Latin America, was included.

Recommendations by the Latin American Energy Organization (OLADE) have been implemented, for instance, in Colombia through the Law 1584 of 2012, which approved a decision of the ministers meeting in the framework of the OLADE. In terms of renewable energies, Ecuador, through policy 4.3 of the National Plan for Good Living[23] includes in 2011, the observatory of renewable energies in Latin America and the Caribbean.[24] In this case, the institutional framework consists of the Ministry of Electricity and Renewable Energy (MEER) as responsible

---

[22] Velázquez Elizarrarás, J. C. (2009). El nuevo derecho internacional de la energía a través del estudio de sus fuentes y el ordenamiento del mercado mundial del petróleo en un contexto geopolítico–especulativo. *Anuario mexicano de derecho internacional*.

[23] Secretaria Nacional de Planificación y Desarrollo. (2009). *Plan Nacional p ara el Buen Vivir 2009–2013: Construyendo un Estado Plurinacional e Intercultural. República del Ecuador, Plan Nacional de Desarrollo.* Quitó, Ecuador.

[24] OLADE. (2011). *Observatorio de energías renovables en América Latina y El Caribe, Ecuador. Informe Final, Producto 1: Línea Base de las Tecnologías Energéticas y Producto 2: Estado del Arte de las Energías Renovables.* ONUD.

for designing and executing renewable energy development programmes. In addition, the National Electricity Council (COMELEC) is in charge of regulating the electricity sector, approving concessions for the use of renewable energy resources, and setting the price of these energies.

Furthermore, in Colombia, the Law 1950 of 2019 ratified the accession of Colombia to the OECD. This type of recommendation usually is followed by most members. This is the case with Mexico in particular with the incorporation of the international tax law measured in the OECD/G20 action plan to correct asymmetries between tax laws of the members. Not following this type of recommendation would result in expulsion from the OECD.

A number of other organisations have contributed important decisions and recommendations. These include The Andean Community on energy matters, the Central American Regional Electric Interconnection Commission (CRIE) and the International Energy Charter, all of which are valuable and important. It should also be mentioned that other international horizontal norms come from agreements between states via bilateral treaties or simplified agreements, and they also are recognised as rule-makers.

We also find the proliferation of private international organisations, traditionally called international Non-Governmental Organisations (NGOs). This type of regulation is called by the doctrine Third Party Regulation.[25] NGOs issue norms that come from international private self-regulation, of which is considered soft law. Examples of such non-governmental energy organisations are World Energy Council, the Initiative of Transparency in Extractive Industries (EITI), etc. It is also worth highlighting organisations such as the Ibero-American Association for Regulation (ASIER), the Ibero-American Association of Energy Law (ASIDE), the Spanish Energy Association (AEDEN).[26] Although these organisations do not issue recommendations, they do have in a certain way a degree of influence as evidenced by the conclusions of their discussions.

---

[25] Barton et al. (n 20).

[26] Vicente López-Ibor, p. o. (n.d.). "The Emergence of a Global Energy Law in the I AEDEN Congress' The Law of Energy in Transition" (C. S. University, & T. e. Newspaper, Interviewers).

The central issue of this entire system of international norms, especially of the norms of international public organisations, whether they are of hard or soft law, is that their incorporation into the internal law of each country can be disrupted. They can be incorporated by direct application or indirectly by a gradual and reciprocal harmonisation. The incorporation depends on the international instrument and how it has established its incorporation into domestic law. The truth is that nowadays, due to the incorporation of global institutions, the energy law of different countries is very similar in their institutions, of course, with certain differentiating nuances. However, the common elements are more than the disagreements. That is why it is even possible to discuss global energy law.

Finally, it should be added that at the international level, a recognition of the human right to energy has been presented at the level of international doctrine, but not in positive law. Parente, (Parente, 2010)[27] defends the assertion that it has been recognised as a human right to energy in the European Union and expressly, in the European Convention to protect human rights and fundamental freedoms. However, in general, this recognition has not been made. What is configured every day is the concept of the right to energy as derived from the field of access to energy. In short, all of this feeds the idea of constructing a global energy law.

2.4 Conclusions

Before 2010, energy law in Latin America and particularly in Colombia was a dispersed and autarchic right. Dispersed because it was composed of separate disciplines and autarchic because it concentrates its interest in determining the regulations only for the organisation of coal, oil, electricity and gas industries, without taking into account other considerations of sustainable development.

Since 2010, energy law has been transformed: first, it became a sustainable energy law due to the introduction of principles based on

[27] Parente (n 15).

the sustainable development goals and measures against climate change. Second, it has transformed into a more unified energy law, due to energy transition. Third, it became a global energy law due to the internationalisation of the regulations.

Part II

Hydrocarbons, Electricity and Mining

3

The Peruvian Renewable Energy Experiment: Lessons for Policymakers

Abel M. Venero Carrasco

3.1 Introduction

Since 2010, Peruvian energy regulation has seen the establishment of regulatory mechanisms to incentivize the development of power generation from non-conventional renewable sources (i.e. wind, solar, biomass and hydroelectric power capacity below 20 MW).[1] In 2010, the country enacted specific legislation in order to integrate non-conventional power

[1] The upper limit of 20 MW set for RER in Peru certainly exceeds the industry's consensus that sets this limit in 10 MW. For example, see Hydro, A. M. E. O. L. "Twelve Reasons to Exclude Large Hydro from Renewables Initiatives" (2003) https://www.rivernet.org/general/hydropower/12reasons.pdf, p. 2. Accessed 18 May 2016, p. 2. Accessed 18 May 2016. The inclusion of the hydropower capacity at this scale was not largely debated in the country, it only deserved a mention in the specific legislation creation Cf SPIJ (n.2).

A. M. V. Carrasco (✉)
Santivanez Abogados, Lima, Peru
e-mail: abel.venero@gmail.com

G. Wood and J. F. Neira-Castro, *From Fossil Fuels to Low Carbon Energy Transition*, Energy, Climate and the Environment,
https://doi.org/10.1007/978-3-031-00299-1_3

generation capacity (named RER for its acronym in Spanish), with the objective to create an incentive programme for the development and commissioning of renewable energy capacity in the market.

This chapter presents a brief introduction to describe the most relevant components of the RER legislation and argues how and why the Peruvian trailblazing approach is a cautionary tale for efforts to incentivize the power generation transition towards RER. Particularly, the Peruvian experience should serve policymakers to consider two specific takeaways: (i) these type of decisions shall be part of a global and integral energy policy approach, specifically the objectives and the metrics to assess success (or failure) of any incentives programme should be specified as part of its enactment; and (ii) the implementation of a market approach incentive without fully incorporating this type of generation into the market dynamics causes regulatory disturbances of which the economic effects are, as usual, borne by the consumers. Whether this effect is worthy will depend on the programme's objectives and success.

Finally, this chapter presents an analysis of the short-term future of the RER generation in the country, as currently the Peruvian Congress is discussing a bill which seeks to update the RER legislation. In this assessment, we find that the lessons learnt from the current iteration of the RER legislation have not been properly considered by the legislators.

3.2 The Legislation and Its Results

3.2.1 Context for the Incentive Programme Enactment

In 2008, as part of the legislative package required for the implementation of the free trade agreement executed between Peru and the United States, the Peruvian Government issue Legislative Decree 1002 that set the pillars for the promotional programme. Despite being enacted as part of the free trade agreement, no specific policy goals or even proper justification for the programme was provided. The objectives set forth for the Legislative Decree enactment report highlighted the need to promote renewable energy sources to reduce hydrocarbons use in the industry and the need to protect the environment, including addressing climate change.[2]

Up until the enactment of the Legislative Decree, the Peruvian potential for hydroelectric capacity had been largely reviewed and, according to these assessments the enormous opportunities for this technology were clear. In fact, the country has a theoretical hydroelectric potential of around 206 GW.[3] Through a methodological process, the report discarded the potential that due to technical reasons cannot be taken. When this is considered, the country's 'realizable' hydroelectric potential is 58.9 GW.[4] A similar assessment regarding RER potential was not, up until that time, properly conducted.[5]

3.2.2 The Main Features of the Legislation

The RER legislation included a market-driven approach to incentivize the implementation of RER capacity. This mechanism was built on three pillars: (i) access of the RER units to the Peruvian energy spot market; (ii) priority dispatch by the system operator (COES) and (iii) eligibility to a guaranteed income contract. These pillars should allow the country to achieve a target RER market share (in MWh) now lower than 5%, excluding hydroelectric capacity.[6]

---

[2] This legislation is accompanied by its bill reports which further describes the legislator's motives. SPIJ. *Exposición de Motivos – Decreto Legislativo 1002* (2008) http://spij.minjus.gob.pe/Textos-PDF/Exposicion_de_Motivos/DL-2008/DL-1002.pdf. Accessed 18 May 2016.

[3] MINEM, *Evaluación del Potencial Hidroeléctrico Nacional*, Vol. 2 (1968) http://www.minem.gob.pe/minem/archivos/file/Electricidad/potencial/Volumen02/PeruPotHidro_vol02.pdf, p. 29. Accessed 30 April 2021.

[4] Ibid., p. 48.

[5] Currently, MINEM has conducted research on the wind power potential in the country. It has published a Wind Power Atlas (2016). http://www.minem.gob.pe/_publicacion.php?idSector=6&idPublicacion=537. In the case of solar power potential, MINEM teamed up with the meteorological public institute Senamhi and published the Peruvian Solar Energy Atlas (2013) http://repositorio.senamhi.gob.pe/handle/20.500.12542/343.

[6] A more detailed analysis of the legislation features can be found in Mitma, R. E. "Análisis de la Regulación de Energías Renovables en el Perú". *Derecho & Sociedad*, 45 (2015), p. 168 http://revistas.pucp.edu.pe/index.php/derechoysociedad/article/view/15235.

Regarding the spot market access and the dispatch, from a commercial and technical point of view, the RER units did not evidence extraordinary treatment. Particularly, the priority dispatch was consistent with the general regime in which COES was required to programme the system operation in order to prioritize generation capacity based on the power production marginal cost.[7] In fact, one can argue that a specific piece of legislation was not really required for the implementation of these parts of the RER capacity incentive programme.

The enactment of specific legislation, however, was needed for the special treatment proposed for the commercialization features set forth in the RER incentive programme. Regarding the new RER capacity, the legislative decree created a particular commercialization scheme in which RER generators could bid a guaranteed price in a biannual bidding processes organized by the energy regulator. In these bidding processes, the regulator signals the required capacity considering the target market share.

The RER capacity retribution considering the guaranteed price is structured as a *contract for difference*,[8] in which the Peruvian government promises the generators awarded in the bidding processes the difference between their earnings in the spot market and a guaranteed price. To comply with the difference guarantee, the RER legislation implements a levy, called the RER premium, to be collected from all energy end consumers. The collection method for this levy is the incorporation of an additional unitary charge as part of the system's wheeling costs.

The collection method for the levy was not new in the country. Prior to the RER legislation, successive Peruvian governments had been enacting similar levies to provide financial feasibility to a series of energy infrastructure projects; the first of these was an additional wheeling charge destined to provide a guaranteed income to the initial natural

---

[7] The Peruvian power sector fundamentals are thoroughly described in Mughal, S. "The Scenario of Restructuring and Deregulation in Peru—Developments and Issues". *International Journal of Electronics and Computers*, 3(1) (2011), p. 75.

[8] Further explanation on the dynamics of this type of future contract can be found in Brown, C., et al. Exchange Traded Contracts for Difference: Design, Pricing, and Effects. *Journal of Futures Markets*, 30(12) (2010), p. 1108. Retrieved from https://citeseerx.ist.psu.edu/viewdoc/download?doi=10.1.1.153.7473&rep=rep1&type=pdf. Accessed 5 November 2019.

gas pipeline system in the beginnings of the 2000s.[9] From that experience on, new levies have been implemented in order to guarantee, for example, the difference between the production costs and the spot market prices, the retribution for emergency generation plants, the development of cold-reserve plants and the expansion of the natural gas pipelines system.[10]

The preference of the government to create this type of levies arises mainly from the reliability of the energy sector payment mechanisms. As part of COES function, the system operation includes the monthly clearance of the spot market in which COES sets forth the compensations due between the generation companies for their production. During this clearance, COES also sets out the retribution amounts for the transmission companies which should have been collected by the generation companies from the end consumers as part of the wheeling costs. Notably, the collection of these costs (and the retribution to the transmission companies) is based on the peak-hour energy demand as the Peruvian capacity market has been designed. The reliability of this clearance arises from the system operator prerogative to suspend a generator in case it fails to timely pay the generation and transmission compensations. In this auspicious environment, the inclusion of additional remunerations to be collected by the generators and transferred via the clearance mechanism benefits from the incentives set forth securing the payments between the energy sector agents.

Finally, the RER legislation left pending the possibility for the non-conventional RER generators to freely trade their capacity through

[9] This mechanism received the name of 'Main Pipeline Guarantee' and served to provide financial relief to the first natural gas transportation pipeline. Further details on this Liceti J., and Gonzalez, G. *Regulando el cambio de la matriz energética del país la regulación del gas natural en el Perú. THEMIS Revista De Derecho* (52) (2006), p. 126. Retrieved from http://revistas.pucp.edu.pe/index.php/themis/article/view/8827. Accessed 5 November 2019.

[10] This is the case of the infamous Southern Pipeline which sought to create a parallel natural gas transportation pipeline. This Project was ultimately suspended and the government contract with the sponsor, the Brazilian company Odebrecht, was terminated. As of today, the government has signalled its future plans to relaunch this Project, however not specific plans have been released to date. The mechanics of this levy is explained in Antayhua, O. J., and Huamán, R. D. P. *¿ Debe reconsiderarse el mecanismo de ingresos garantizados para el gasoducto sur peruano? opciones de viabilización del proyecto y análisis de mercado* (2018), p. 68. Retrieved from https://repositorio.up.edu.pe/handle/11354/2130. Accessed 30 April 2021.

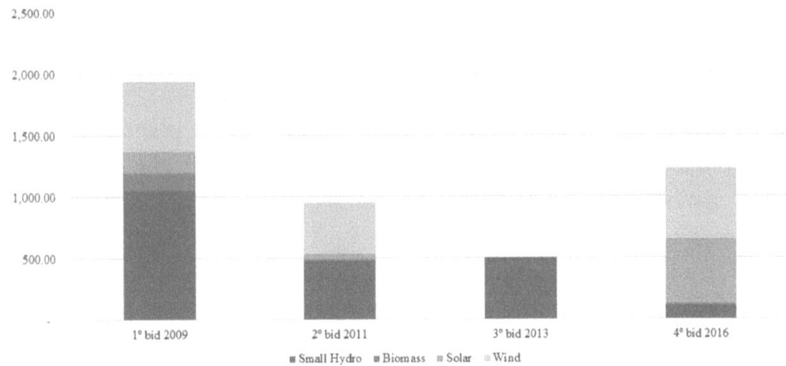

Fig. 3.1 Annual energy (GWh) awarded by technology (2009–2016)

PPAs but did not put in place specific rules. In particular, for PPA trading, the generators are granted with 'firm capacity' recognition which reflects their generation units' reliability and, using this trading capacity, the generators are allowed to execute PPAs with end consumers. As mentioned previously, the RER legislation did not set specific rules, partially because, at the time it was enacted, RER PPA trading was not really viewed as feasible.[11]

3.2.3 The Bidding Results

As of today (autumn 2021), the energy regulator has conducted four bidding processes that have resulted in several RER contracts being executed between the Peruvian State and the generation projects developers. Figure 3.1 presents the outcome from these bidding processes.[12]

Several conclusions can be drawn, in particular regarding the success of the bidding process to provide new capacity and cover the capacity

[11] As we will show in the next paragraphs, RER market prices were way above the energy trading prices in the spot market, so PPA trading was not feasible and not proper attention was given to the topic.

[12] The figure is made by the author using OSINERGMIN information provided as part of the RER remuneration for the period May 2021 and April 2022. https://www2.osinergmin.gob.pe/GRT/Procesos-Regulatorios/Tarifas-Barra/FPB-2021-2022/09PrecioBarra/9.2.5.CargosAdicionales.rar.

requirements set forth as part of the bidding process (in general, the bids achieved 99% of the auctioned capacity).[13] In hindsight, the results have barely achieved the market share target: for 2020, with several restrictions due to the COVID-19 pandemic, the RER market share reached 5.25%.[14] Apart from the market share indicator, the economic pledges that the Peruvian State has made regarding the guaranteed prices must be taken into account: annually US$297 million have been promised for the following 15 years. For the year 2021, these pledged amounts will require the collection of US$215 million in levies from the end consumer[15]; around 80% of the RER remuneration comes from subsidies to the energy users.

Do these results prove that the promotion mechanism was sound energy policy? Are the economic pledges forced upon the energy consumers fair and reasonable? Unfortunately, we cannot properly answer these questions because, as mentioned above, the policy was put in place without clear objectives in the first place. The RER incentive programme was enacted for the sake of promoting non-conventional technologies, but no specific goals were stated and proper metrics to assess the programme were not included anywhere. How could we assess if the legislation has been successful (or if it has been a failure) if we do not know what success looks like?

Some may argue that the legislation has helped Peru curb its greenhouse gases emissions and that by itself ratifies that it has been good policy. This is not a wrong approach but, as in the case of the legislation, this approach to the assessment is incomplete as one cannot answer, for example, if this reduction in emissions is sufficient for the country to comply with its Paris Agreement pledge.[16] More importantly, we are also left with the questions of whether these costs were effective and what one could expect from the market.

---

[13] Vásquez, Arturo (Ed.). *La industria de la energía renovable en el Perú: 10 años de contribuciones a la mitigación del cambio climático* (2017), p. 106. Retrieved from https://www.osinergmin.gob.pe/seccion/centro_documental/Institucional/Estudios_Economicos/Libros/Osinergmin-Energia-Renovable-Peru-10anios.pdf. Accessed 30 April 2021.

[14] The market share information has been retrieved from COES 2020 statistics https://www.coes.org.pe/Portal/areas/Publicaciones/documentos/2020/capitulo04.htm.

[15] Information provided by OSINERGMIN (n.15).

[16] This approach is discussed further in Sect. 3.4.1.

If a lesson has to be drawn from these results, and in general from the Peruvian experience, it is that policy decisions have to be implemented including the specific goals that they are set to achieve. Failure to provide a signal in terms of what a specific policy seeks to achieve is an undesirable situation and should be avoided. Policymakers seeking to incentivize the transition to systems with significant shares of renewable energy should not only focus on the policy structure and the rules to be put in place accordingly, but they also need to specify what the policy ultimately seeks to achieve. Otherwise, as in the case of Peru, we are left with a big question mark regarding the necessity of the RER legislation and, furthermore, whether its rules require further amendments.

In our opinion, good and sound policymaking requires the proposals to clearly state their drivers and the intended consequences, otherwise the specific policies cannot be properly assessed. This is particularly relevant in developing countries in which the resources are by definition scarce and efforts should be placed where they are expected to be fruitful.

Beyond this lesson, a more complex one arises when the results are assessed from a wider point of view. In particular, the Peruvian approach for the inclusion of new RER has caused several undesirable effects in the energy market which, in turn, enlarges the question mark on the RER legislation in terms of necessity and even its quality as good legislation.

3.3 Effects on the Power Market

3.3.1 The Relevant Commercialization Rules

From the discussions above, the Peruvian market has been built using a pool system approach in which the generators participate in a spot market where they clear their production against the supply commitments they may have executed with their clients.[17] The generators' clients are the distribution companies obligated to supply to all the regulated clients within their concession zones and individual unregulated clients.

The end consumers' classification between regulated and unregulated clients depends exclusively on their power capacity demand. An

[17] This approach to the bulk and wholesale system is described in Mughal (n.7).

unregulated client is the power consumer demanding 2.5 MW or above, while the regulated clients are those with capacity demands below 0.2 MW. Between these limits (0.2 to 2.5 MW), the regulation provides the right of the power consumer to elect their classification as either regulated or unregulated.[18]

The unregulated clients can freely negotiate their PPA terms and conditions except for the regulated conditions for the application of the wheeling costs. For the case of the regulated clients, as explained, these are mandatory end consumers supplied by the distribution companies which are subject to specific rules for their energy procurement.

In the case of the distribution companies, they procure energy from the generation companies using two methods. First, these companies can agree a bilateral freely negotiated contract subject to a price cap determined by the energy regulator of Peru (in Spanish *Organismo Supervisor de la Inversión en Energía y Minería*) (Osinergmin) on a yearly basis; on the other hand, they can require the energy regulator to conduct a public bidding process to award long-term PPAs (up to 20 years) with prices higher than the price cap.[19]

By the time the RER legislation was enacted, the industry was amid the implementation of the initial bidding processes. Under these contracts, mostly all of the energy requirements for the regulated clients were secured under PPAs between the generators and the distribution companies.[20] These PPAs have take-or-pay capacity provisions and also mandatory long-term duration with no unilateral remedies for the distribution companies in case of part of the clients leave the regulated condition.

These were the main market rules and, as explained above, came under strain partly due to the RER legislation regarding the levies that serve to guarantee the income promised to the RER generators.

---

[18] Art. 3. Supreme Decree Nº 022-2009-EM - *Reglamento de Usuarios Libres* published 16 April 2009.

[19] Art. 3. Law 28832 - *Ley para asegurar el desarrollo eficiente de la generación eléctrica* published 23 July 2006.

[20] According to the information used by OSINERGMIN to calculate the end consumer tariffs, for December 2020, the bilateral contracts amounted only 5% of the supply commitments for the regulated client's supply. The rest is allocated in auctioned contracts with mandatory terms and no remedies in case clients leave the distribution company exclusivity.

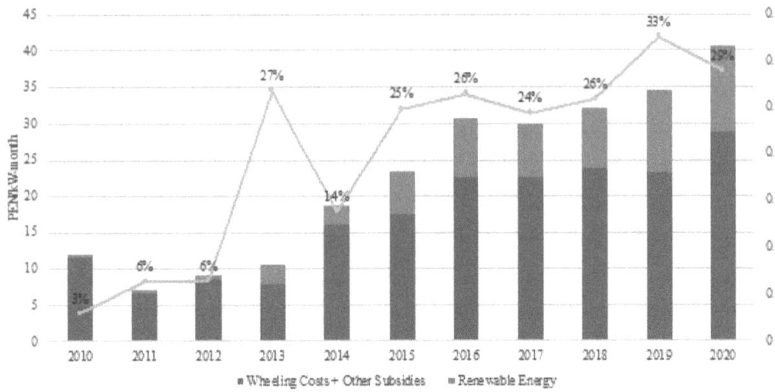

Fig. 3.2 Wheeling costs and RER share increase

3.3.2 The Intrinsic Problem of the RER Levies

Conceptually, the levy mechanism presents an intrinsic problem that was not adequately assessed prior to the enactment of the RER legislation. With an increasing market share of RER, the marginal costs were driven lower and lower as they have minimal variable operation costs. In fact, although the new renewable capacity was not the only issue at blame,[21] by themselves the levies were self-defeating from a consumer point of view: the more successful the inclusion of these new technologies, the lower the spot market prices and the higher the levies and the wheeling costs required to fulfil the Peruvian State pledges. In fact, Fig. 3.2 shows the growing share of the RER levies as part of the wheeling costs.[22]

The end consumers have been funding the development of new generation capacity which represents cheaper sources of energy production, but their prices have been ever-increasing. In theory, the increase should

[21] This situation has also been caused by the government intervention in order to expand the generation capacity. By the beginning of the 10's decade, the government promoted several new generation projects in order to supply the total demand which was expected to further increase. The demand growth rates were lesser than expected, thus there was an overcapacity situation in the system. Supreme Decree N° 022-2018-EM - *Reglamento de Usuarios Libres* published 5 September 2009 explains the mismatch between the government expectations and the real growth.

[22] The figure is made by the author using OSINERGMIN (n. 12).

have not been problematic as the levies were thought to distribute the cost among all the end consumers.

The RER legislation, however, did not foresee that the self-defeating effect of the RER technology (i.e. drive the market prices lower) was going to be unevenly passed through to the consumers, affecting mainly the distribution companies' regulated clients. As explained, supply for these regulated clients was secured under long-term PPAs which lacked price adjustment mechanisms. The continuous increase in the wheeling costs as seen in Fig. 3.2 also drove up overall energy costs for these clients. Under this circumstance, the market response was, as many would expect, to seek the opportunities to optimize their energy costs.

3.3.3 The Market Response

As the energy prices for the regulated clients were increasing, many clients foresaw the opportunity to change their energy supply regime and access the unregulated client category. This change benefits them as they can now hire lower market prices created as a pass through of the lower spot market prices (average unregulated prices went from 186.9 PEN/MWh[23] in 2015 to 150.3 PEN/MWh).[24] The share of these clients rapidly increased from 320 in 2015 (with energy purchases equal to 18,262 GWh or 46% of the total energy sales)[25] to 2,248 in 2019 (purchases equal to 28,132 GWh or 60% of the total energy sales).[26]

As unregulated clients, the 'migrating' users were eligible to execute independent PPA with the generators, which, in turn, passed through their lower energy costs in the spot market. These new unregulated clients saw their energy prices reduced simply by choosing to change

[23] OSINERGMIN. 2015 Unregulated Market Bulletin, p. 32. Retrieved from https://www2.osinergmin.gob.pe/Publicaciones/pdf/BolMerLibElectricidad/MERCLIB2015.pdf.

[24] OSINERGMIN. 2019 Unregulated Market Bulletin, p. 29. Retrieved from https://www2.osinergmin.gob.pe/Publicaciones/pdf/BolMerLibElectricidad/MERCLIB2019.pdf.

[25] OSINERGMIN. 2015 Annual Statistics, p. 77. Retrieved from https://www2.osinergmin.gob.pe/Publicaciones/pdf/Anuario/Anuario2015.pdf.

[26] OSINERGMIN. 2019 Annual Statistics, p. 75. Retrieved from https://www2.osinergmin.gob.pe/Publicaciones/pdf/Anuario/Anuario2019.pdf.

their category as clients, however this was not the only price reduction opportunity.

As explained previously, the levies collection method relied on the wheeling costs, which are collected considering the peak hours capacity requirements. Being unregulated clients, many of these were able to free themselves partly from these levies by avoiding peak hours consumption whether via their internal reorganization or by the implementation of peak shaving generation units.

As companies reduced their wheeling costs, the levies collection also fell; thus, the unitary levy increased in order to compensate. In turn, these increases incentivized more clients to change to the unregulated market and avoid the levy collection with the effect of new increases in the unitary RER levy.

In this vicious circle, the ones that see their energy costs continuously increasing are the regulated clients that do not have the chance to free themselves from their regulated conditions. These are mainly the residential consumers. Here is where the policy should be ultimately judged: has the achieved goals of the RER legislation (which were not explicitly stated) worth the burden posed on the regulated clients?

3.3.4 The Future of the Market

Since 2016 no new RER bidding processes have been conducted, however it is not clear if this has been a damage control decision or simply the change in governmental priorities. As no new RER capacity is commissioned in the market and the natural demand increase, the marginal costs should reverse course and start to be driven upwards. In this situation, the incentives for the clients to respond to an ever-growing energy price might disappear and many of them would be prompted to return to the regulated market as the generators will seek to pass through higher marginal costs.

As other market forces and regulatory developments are expected to drive the marginal costs higher[27] despite the RER market share, the Peruvian renewable policy experiment has shown that the half-hearted market approach was defective. Furthermore, the clients that ended paying the bills were the smallest ones. Clearly, this is not what good and sound policy looks like.

3.4 The Future for RER Legislation in Peru

Supporters of renewable energy have been arguing that the current market-based approach is not completely flawed and that it requires only a minor tweak. In particular, they argue that development of new capacity will be feasible if RER units are fully incorporated into the energy market by means of allowing these units' energy trading through PPA.[28] As explained previously, this is a feature not included initially as the RER legislation did not provide specific rules for the recognition of 'firm capacity' to these units.

Renewable energy promoters imply that the failure of the regulatory framework to recognize firm capacity to the RER units has been the major obstacle for the success of the incentive programme. Recently, the Peruvian Congress discussed (but was subsequently rejected on 5

[27] This mainly refers to the regulatory change ordered by the Peruvian Supreme Court which requires MINEM to provide equal treat to the natural gas fired units when their marginal costs are calculated. Prior to this decision, these units' marginal costs were determined using the information declared by the generators while all other technologies' marginal costs were calculated using actual information. As natural gas generators declared zero costs for their production (as they included natural gas take-or-pay provision as fixed costs), their marginal costs, and accordingly the overall system cos, was driven low. Further details of this case can be found in Yesquen, K. *Los precios de gas natural en la generación eléctrica: ¿un nuevo capítulo en esta historia?* (2021) https://rubio.pe/publicacionescont/los-precios-de-gas-natural-en-la-generacion-electrica-un-nuevo-capitulo-en-esta-historia/. Accessed 5 May 2021.

[28] As of today, there has been some advancements in the 'firm capacity recognition', however there are pending issues to discuss. For the details on the discussion, see Aguirre, M. *La potencia firme para las renovables, ¿va o no va?* https://revistaenergiaynegocios.com/2019/09/23/la-potencia-firme-para-las-renovables-va-o-no-va/. Accessed 30 April 2021.

May 2021)[29] a bill to expand the initial 5% market share target to a 20% objective in 2030 and 50% goal in 2050,[30] while setting specific rules to recognize tradable capacity to the non-conventional plants. This piece of prospective legislation, additionally, requires the government to reinitiate the bidding processes, providing a short-term timetable for the energy regulator to call for new bidding processes. The conclusions from the previous sections of this chapter show that the framework may need several adjustments prior to its enactment.

3.4.1 Making the Same Policymaking Mistake

The new proposals are again on the wrong side of policymaking as they not only ignore the current situation of renewable energy generation in the country, but it also sets a goal without really considering the situation of the energy market. In particular, the bill specifies that it seeks to contribute to the goal to reduce Peruvian greenhouse gas emissions, however this goal is set far from the technical consensus and even far from the assessment made by the competent bodies regarding the Peruvian climate change response.

As part of the implementation of the climate change emission reductions pledge from the Peruvian Government under the Paris Agreement, the technical advisors comprised of representatives from all the economic sectors have stated that, in order to achieve the Peruvian pledge, the RER target share should be 6% by 2030.[31] Part of this analysis has taken into account the situation of the energy market (which, as explained, has been negatively impacted by the RER approach currently in place) but also

---

29 The renewable energy generators association has considered this as a setback and a harmful decision to all the energy consumers. See https://revistaenergia.pe/electricidad/spr-no-aprobar-la-ley-de-inversion-en-energias-renovables-perjudica-a-todos-los-peruanos/. Accessed 9 May 2021.

30 The bill can be found at https://leyes.congreso.gob.pe/Documentos/2016_2021/Proyectos_de_Ley_y_de_Resoluciones_Legislativas/PL06953-20200118.pdf.

31 GTM-NDC. *Informe Final del GTM - NDC* (2018), p. 61 – Annex 5-2. https://www.minam.gob.pe/cambioclimatico/wp-content/uploads/sites/127/2019/01/190107_Informe-final-GTM-NDC_v17dic18.pdfPA%c3%91OL.pdf.

The annex text can be accessed at https://drive.google.com/file/d/14w8X4szF2-uXPdFSPBeuxtUW1cfXbFlL/view?usp=sharing. Accessed on 30 April 2021.

other cost-effective opportunities in other economic sectors to reduce these emissions (particularly in transportation and industry).[32]

Accordingly, as it requires an increase in the RER share, the bill mandates the regulator to conduct new bidding processes to award new 'contracts for differences' and enlarge the Peruvian State pledges to new generation capacity. Clearly, legislators are not aware that the current approach has proven burdensome for the energy sector and that the regulation based on levies has not been a good mechanism, considering the negative effects it has caused in the regulated clients' market.

3.4.2 Does a Market-Approached Mechanism Rightly Incentivize New RER Capacity?

The new proposal aims to further incorporate the RER capacity recognizing it as tradable capacity in the market. Furthermore, the legislation provides detailed rules for the recognition of tradable capacity for wind and solar plants; in particular for the case of wind generation, the bill exposes the need to recognize tradable capacity equal to the average output of the unit during the system's peak hours. On the solar generation side, the bill proposes their firm capacity to be equal to the lowest capacity output in the preceding three years.

What the previous experience has shown is that a half-hearted market approach can create unintended consequences. Despite the agents' interactions being highly decentralized, the commercial policies coincide with the search to maximize profits and, in that sense, the companies will seek to exploit any advantages the market rules may. The case for market mechanism, then, requires to fully embrace the market dynamics as part of the incentives programme.

Keeping this in mind, it is good and sound policy to seek RER participation in the market by allowing their trade; however, it must be noted that this approach may not be the best. Tradability in the Peruvian system derives from the security the unit provides to fully supply its capacity during the whole day; thus, conventional plants such as hydroelectric

---

32 Ibid., 689.

or thermoelectric have specific rules for the recognition of their 'firm capacity' based on their capacity to reliably produce energy. These rules have been amended continuously and incorporating the lessons learnt from the interaction of these technologies by COES. In none of these cases has its 'tradability' been enacted as part of the legislation justly for the need to further adjust their parameters to better fit them in the COES dispatch.

As opinioned here, this should be the correct approach to non-conventional renewable energy *firm capacity*, setting rules in legislation may prove counterproductive as the market evolves and new interactions between technologies are observed. This is especially the case given the insignificant share of RER production in the market; the Peruvian technical community has not yet fully comprehended the effects of the intermittency of the RER sources which will challenge the current approach to capacity trading in the system. We may anticipate that this specific topic may be the future cause of market disturbances as current tradability is understood as 24-hour production reliability.

3.5 Conclusions

As obvious as it may sound, no rule in the energy sector (or in any other sector) works in isolation to the rest, even in cases where it is clearly and expressly built to work as a stand-alone provision. Thus, the decisions of policymakers to increase the renewable energy share in their power markets require such initiatives to be assessed not only from the perspective of the individual goals, but also in the context of the energy sector as a whole. In the specific case of the Peruvian legislation related to renewable energy, unfortunately, both assessments do not have a good outcome.

From a stand-alone policy perspective, a proper assessment cannot even be performed, as the RER legislation was enacted without specific goals; at times it seems the legislation sought to incentivize renewable energy generation for its own sake. This should serve new policymakers in Peru (and in other countries) to bear in mind that their policies require not only the construction of a coherent set of rules for its implementation, but also the need to have proper objectives that will allow revision of the policy if it has not been adequately implemented and whether the rules should be amended.

From a point of view that considers the place of the legislation in the grand scheme of the energy market, RER legislation has proven damaging to the end consumer. An assessment of these results should lead Peruvian policymakers to reject the current approach and build different policies instead of the promotion of new renewable energy generation using the 'contracts for differences' tool.

Current developments in the regulatory environment, however, show that policymakers, particularly legislators, are not aware of the results from the RER legislation as they insist on the current mechanism that has proven to be significantly burdensome to the end consumer without a clearly defined beneficial goal.

Policymakers in the country and other countries should view the Peruvian experience as a cautionary tale and take-away the need that this type of decisions shall have clear objectives in order to be assessed in the future. Additionally, they should also take into account that the implementation of market approach incentives for renewable generation without fully incorporating them into the market dynamics causes disturbances that are ultimately borne by the end consumers.

4

Searching for Security of Energy Supply Through Renewable Energy Sources: A Practical Overview of the Chilean Case

Andrés Zamorano

4.1 Introduction

The main objective of this chapter is to provide an overview of the progress status of the adoption and later execution of a set of short-, medium-, and long-term public policies to promote the diversification of the power energy matrix through electrical projects based on Non-Conventional Renewable Energies (NCRE), in search of security of energy supply for Chile. Making visible the intense process of regulatory changes that began in 2004, leading to strengthening the entire energy infrastructure in its broadest sense (regulatory, administrative, structural, and technical, among others), may eventually serve as a precedent for those countries or economies that are engaged in similar energy matrix transitional processes in terms of promoting the development of NCRE.

A. Zamorano (✉)
Private practice in Administrative and Regulatory Law, Santiago, Chile
e-mail: zamorano.andres@gmail.com

G. Wood and J. F. Neira-Castro, *From Fossil Fuels to Low Carbon Energy Transition*, Energy, Climate and the Environment,
https://doi.org/10.1007/978-3-031-00299-1_4

By taking due consideration of any of the multiple factors that promote their development, either for environmental awareness, to secure lower production costs at the industrial level, or due to the need to achieve independence in access to energy, as in the case of Chile, other countries can assimilate their own processes with those under development in Chile.

Initially, we must understand the historical scenario related to the access to fuels, as well as the events that led successive Chilean governments to seek a transformation of the energy matrix through change and adjustment of the energy policy framework.

Visualising this scenario, eventually we will be able to glimpse the contribution made by the implementation and development of a comprehensive energy policy framework for the development of native energies based on NCRE, capable of granting security of energy supply and in this way promoting economic and social well-being.

It is particularly noteworthy that without direct subsidies to investors, electricity generation from solar photovoltaic, wind sources and other NCRE technologies and fuels have been responsible for completely transforming the Chilean electricity generation system. The main reason is geography and the abundance of resources (solar radiation and wind) and the progressive reduction of technology costs and regulatory adjustments.

In an attempt to address the historical problems relating to the availability of reliable energy sources, we will explore the dimensions of energy supply security in the context of Chile by setting out the scenario prior to 2008 and the context in which the issues will be considered. The next section provides an overview of Chile's regulatory changes, and in doing so considers more closely the key energy issues and the rationale behind the search for long-term stability and access to energy. Subsequently, this chapter offers a look at the most relevant regulatory changes enacted as part of the pool of initiatives adopted to provide a long-term energy policy. Finally, this chapter provides a number of projections related to the availability of new energy and the opportunities envisioned as the next steps to come.

4.2 Historical Domestic Facts: Chile

4.2.1 Domestic Overview

The Republic of Chile is a South American country occupying a long, narrow strip of land. It runs 4,300 km from North to South but only 175 km from East to West, between the Andes mountain range to the East and the Pacific Ocean to the West. It borders Peru to the North, Bolivia to the Northeast, Argentina to the East, and the Antarctic in the far South.[1] Chile is geographically isolated and does not have access to reliable domestic reserves of cheap fossil fuel resources[2] to support the growth and economic wealth of its people.[3] Regarding energy policies, Chile was the first country to liberalise the electricity sector via privatisation of the state-owned utilities, a process that began with the General Law of Electric Services (LGSE) originally enacted in 1982, and later executed between 1986 and 1998.[4]

The country is, however, the world's largest copper producer, representing about 9.1% of total global production. Chile is also a large mineral producer worldwide. In terms of domestic growth, the mining industry has historically represented about the 10% of the Gross Domestic Product (GDP), about 18% of total foreign investment, and around 52% of total exports and foreign exchange earnings of the country.[5]

By understanding that mining is the country's economic engine, the whole mining industry, and the country itself therefore depends on the availability of cheap energy to remain competitive.

It is important to remark here that the energy-intensive mining sector consumed about 25.5 TWh in 2020, representing more than 33% of

---

[1] Library of Chilean National Congress: https://www.bcn.cl/siit/nuestropais/index_html.

[2] Chile's oil record production accounted for only 3% of its total oil consumption in 2012 (around 1 thousand barrels per day [kb/d]).

[3] IEA, 2014. *Energy Supply Security: Emergency Response of IEA Countries 2014*, s.l.: Astrid Dumond, Therese Walsh, Barbara Zatlokal.

[4] IEA, 2018. *Energy Policies Beyond IEA Countries - Chile 2018*, s.l.: IEA Publications.

[5] Mining Council, 2021. *Updated Mining Figures*, Santiago: Chilean Mining Council.

total national electricity consumption. Furthermore, according to official reports about 12% of the mining operational expenses are directly related to the cost of electricity used in production processes.[6] This is the basic scenario in need of comprehension; the demand of the country, accounting for strategic long-term energy policies aimed at achieving security of energy supply and providing consistent, sustainable economic growth.

4.2.2 Energy Availability

The Chilean energy sector has always been identified as unique compared to its fossil fuel rich neighbours. Chile has no oil, gas, or coal (at least in any relevant amounts). As such, it is a net importer of fossil fuels, and the energy production has been traditionally anchored to hydropower generation and fossil fuel-based systems.

As pointed out by the International Energy Agency (IEA), Chile is a large net importer of hydrocarbons (crude oil, oil products, and Liquefied Natural Gas (LNG), among others), mostly used in thermal power generation plants to complement or replace hydropower generation according to season or geographical location.[7]

These characteristics have shaped energy policy over time, especially the issue of security of energy supply.[8] In Chilean terms this is focused on, among other factors, power generation and its availability at all levels, associated energy policies such as diversifying fuel supply, generation technology mix, and environmental concerns.[9] In terms of planning and the regulation of energy-related initiatives, it is possible to identify two discrete periods regarding energy policy.

The first period between the years 1982–2008, within the context of an open economy and liberalised electricity market, was clearly related

[6] Ibid.

[7] Ibid., 3.

[8] As stated by the Commission of the European Communities the concept of Security of Supply can be understood as the 'uninterrupted physical availability of energy products on the market, at a price which is affordable for all consumers (private and industrial)'; 'Towards a European strategy for the security of energy supply', Brussels (29.11.2000), COM(2000) 769 final.

[9] Energy Ministry, 2015. *Energy Policy for Chile: Energy 2050*, Santiago: s.n.

to developments driven by the private sector and business expectations. During this period, the electricity sector was mostly managed according to the needs of the mining industry and private initiatives. As an illustrative example, in 2008 Chile's power sector had an energy matrix from primary sources integrated by a mix of thermal technology (accounting for 60.1%) and hydroelectric (accounting for 39.9%) of total power generation; this represented 11,545 megawatts (MW) of total installed power generation capacity.[10]

It is possible to identify a second period starting in 2008. This was driven by two important factors; a long-term drought seriously affecting hydroelectric power generation and, Argentina's decision—as sole supplier—to restrict the supply of natural gas to the country.[11] This was essentially a time in which the government as policymaker undertook the essential role of guiding the evolution of the energy sector via pursuing an approach that guaranteed energy security for the territory. This approach assumed a proactive position and sought a strong set of tools and policies to pursue long-term security of energy supply while providing consistent and sustainable economic growth.

4.2.3 Arrival of Non-Conventional Renewable Energy Regulation in Chile

Visualising the need to introduce deep reforms on energy regulation to attract investment, knowledge, and technology, the government, under IEA recommendations, started to work on long-term regulation to transform the regulatory framework that had been virtually frozen since 1982.[12]

As the first attempt to open a door to technology and investment, it is pertinent to mention Law 19,940,[13] enacted in 2004, guaranteeing all small power producers (<9 MW) the right to sell their energy to the

---

[10] CNE, 2016. *2015 Energy Statistical Yearbook Chile*, Santiago: s.n.

[11] In 2005, natural gas accounted for 22% of Chile's total domestic installed power generation (CNE, 2016).

[12] IEA, 2009. *Chile Energy Policy Review 2009*, Paris: s.n.

[13] See in Spanish: http://www.leychile.cl/Navegar?idNorma=222380.

market at spot or nodal prices. In addition, non-discriminatory access conditions to the distribution system was guaranteed. Progressive toll trunk charge exemptions were set up for NCRE producers, from full exemption for less than 9 MW capacity plants, to no exemption for plants with a capacity over 20 MW.[14]

In December 2005, the 'National Energy Efficiency Program' (PPEE in Spanish) was created. The PPEE was the first and main mechanism through which the government's energy efficiency policy and standards have been developed and implemented.

In 2008, the Law 20,257,[15] better known as 'Non-Conventional Renewable Energy Law', was enacted. This law aimed to promote the diversification of the energy matrix by developing NCRE sources, such as geothermal, wind, solar, tidal, biomass, and small hydroelectric. One of the most outstanding features of this law was that it obligated power distribution companies to include at least 5% from NCRE sources in their power capacity portfolio during the period 2010–2014. From 2015 onwards, the aforementioned obligation would show a steady annual increase of 0.5%, progressively reaching 10% of the portfolio by 2024.

It is noteworthy in regards the issue of energy governance that the Ministry of Energy was created in 2009 under Law 20,402.[16] The Ministry is responsible for the definition of macro policies, developing and coordinating plans, policies, and standards for the development of the energy sector.

In 2012, the government released the 'National Energy Strategy 2012–2030',[17] again following the recommendations made by the IEA in its 'Energy Policy Review 2009'.[18] The plan, known as 'Energy for the

[14] Nasirov, S. A. C. S. C. e. a., 2017. Renewable energy transition: a market-driven solution for the energy and environmental concerns in Chile. *Clean Technologies and Environmental Policy*, October.

[15] See in Spanish: http://www.leychile.cl/Navegar?idNorma=270212&idParte=0.

[16] See in Spanish: http://www.leychile.cl/Navegar?idNorma=1008692&idParte=0.

[17] Energy Ministry, 2012. *National Energy Strategy 2012–2030,* Santiago: s.n. See in Spanish: 3_Estrategia-Nacional-de-Energia-2012-2030_Energia-para-el-Futuro.pdf.

[18] IEA (n 12).

Future', was based on six priority areas, these were emphasised by policymakers in order to make Chile's electricity sector cleaner, more secure and more cost-effective in the long term.

The six priorities that the National Energy Strategy focused on are as follows:

i. Energy Efficiency to reduce consumption and decouple growth and energy demand;
ii. Non-Conventional Renewable Energy[19] sources, seeking to diversify the electricity (power) matrix;
iii. Strengthen traditional energy sources such as hydro and biomass;
iv. Strengthen the design of the electricity transmission system towards that of a Public Electricity Highway;
v. Address the challenges presented by the market promoting a more competitive electricity market; and
vi. Promote regional sustained progress via developing international electricity interconnections.

In 2013, the Law 20,698[20] was enacted. This law promotes the expansion of the energy matrix, through the development and deployment of NCRE. In response to the success in generating power from NCRE, this law makes changes to the quotas set by its predecessor (Law 20,257-2008). For example, it increased the requirements on distribution companies to step up the amount of energy produced by NCRE. The growth defined in that legal body established that by 2025, the distribution companies' portfolio must be made up of at least 20% NCRE.

In 2015, the Law 20,805[21] was enacted. This law seeks to improve competition in the electricity bidding process for customers subject to price regulation.[22] It has improved the competitiveness of renewable

---

[19] In Chile, the concept of 'Non-Conventional Renewable Energy' excludes large hydropower projects from the category, which is conceptualised as a Renewable Energy.

[20] See in Spanish: https://www.bcn.cl/leychile/navegar?idNorma=1055402&idParte=.

[21] See in Spanish: https://www.bcn.cl/leychile/navegar?idNorma=1074277.

[22] Pérez-Arriaga, I. J., 2013. *Regulation of the Power Sector.* Pérez-Arriaga, Ignacio J. ed. s.l.:Springer-Verlag London.

energy technologies, thanks to the establishment of preferential time blocks and seasonality for certain technologies.

In 2016, Law 20,936[23] was enacted. This law establishes a new system for the constitution of electricity generation companies resident in Chile, electricity transmission systems, transmission rates and remuneration, long-term planning of transmission expansion, greater facilities for the connection of new generation projects, and the noteworthy creation of a new National Electricity Coordinator (CEN). CEN is independent from the generation companies and is in charge of coordinating the operation of all facilities regarding the National Electricity System. Its operations focus on the following areas: preserving the security of service in the electricity system; guaranteeing the most economical operation for all installations of the electrical system; and guaranteeing open access to all transmission systems. It is active in formulating operation and maintenance programmes and permanently monitors existing competition conditions in the electricity market with the aim of detecting signs of actions that could become constitutive of attacks against free competition. Additionally, CEN prepares the Annual Transmission Expansion Plan, an investment proposal for the different transmission segments, among many others.

In January 2015, the document 'Energy Policy for Chile: Energy 2050' was released.[24] This was intended as a long-term energy policy extending to the year 2050. This new guideline constitutes a major development and was born from a Ministry of Energy initiative tasked with seeking an open participatory planning process aimed at achieving a steady long-term energy policy for the country.

Its goal establishes that by 2050, 'the country has a robust and resilient energy system, can manages risks, allowing it to face and anticipate the effects of energy crises, natural disasters, climatic events and geopolitical conflicts',[25] among other challenges.

The conception of this process considered several levels of discussion and participation including an Advisory Committee, a series of working

[23] See in Spanish: https://www.bcn.cl/leychile/navegar?idNorma=1092695.

[24] Energy Ministry (n 9). See in Spanish: http://biblioteca.digital.gob.cl/handle/123456789/611.

[25] Energy Ministry (n 5).

groups, regional workshops, and a virtual platform to convene broad citizen participation. All of this was done to build a shared vision for the future development of the energy sector with social, political, and technical validation in order to become the energy policy that Chile needs.

This policy as the main route guide in energy development is based on five pillars:

1. Security and Quality of Supply: (a) Security and flexibility at the centralised production level; and (b) Decentralised production and active demand Management.
2. Energy as the engine of development (this focuses on): (a) Inclusive energy development; (b) Equal access to energy services and quality of life; (c) Territorial inclusivity; and (d) Competitiveness in the energy sector.
3. Environmentally compatible energy sources (they focus on): (a) A renewable energy matrix; (b) Local externalities; and (c) Energy and climate change.
4. Energy efficiency and education: (a) Energy efficiency; and (b) Energy education and culture.

In 2018, the Law 21.118 was enacted.[26] This law establishes the possibility that end users, subject to fixed pricing, who have for their own consumption equipment for the generation of electricity by non-conventional renewable technologies or efficient cogeneration facilities, individually or collectively, will have the right to inject the energy generated into the distribution network.

As seen above, through the search for security of energy supply solutions, Chile's power regulation underwent a fast and broad development. It focused on the governance of the electricity market, integrated renewable energies to the electricity mix, allowed small power producers to participate in the spot market, and regulated the market in terms

[26] See in Spanish: https://www.bcn.cl/leychile/navegar?idNorma=1125560&idParte=9968877.

of requiring electricity distribution companies to include a quota of electricity generated from NCRE technologies.[27]

4.3 Non-Conventional Renewable Energy Development in Chile

The complex and comprehensive process of seeking security of energy supply through the implementation of long-term energy policies, promoting energy matrix diversification and advancement of energy transition through NCRE, has given the power sector the chance to develop native energy sources.

Chile's aggregated renewable power generation as of March 2021 represents 25% of the total electrical matrix, equivalent to 1,642 GWh. According to the Chilean Association of Renewable Energies and Storage (ACERA), the installed generation capacity from NCRE sources is about 8,773 MW, representing 31.1% of the electricity pool (See Fig. 4.1).[28] In practical terms, this means that they have already complied with the goal set by Law 20,698 establishing that by 2025, distribution companies' portfolios must be at least 20% from NCRE sources.

Regarding the installed generation capacity from NCRE sources, the National Energy Commission (CNE), in its monthly report on energy developments, stated that for the month of February 2021, the power injected into the grid from solar technology was about (549 GWh), from wind (549 GWh), from Mini Hydro (207 GWh), from Biomass (95 GWh), and from geothermal technology (22 GWh).[29]

Likewise, the CNE indicates that in February 2021, 144 new electricity generation facilities were under construction, representing a total of 5,339 MW. These facilities will be operational between 2021 and February 2023. Regarding the Net Billing segment up to February 2021,

---

[27] For a complete review to the regulatory energy framework in Spanish, access the following link: https://energia.gob.cl/mapa-del-sector-energetico.

[28] ACERA, 2021. *Statistics Generation Sector Renewable Electrical Energy*, s.l.: s.n.

[29] CNE, 2021. *Energy Sector Report March 2021*, Santiago: s.n.

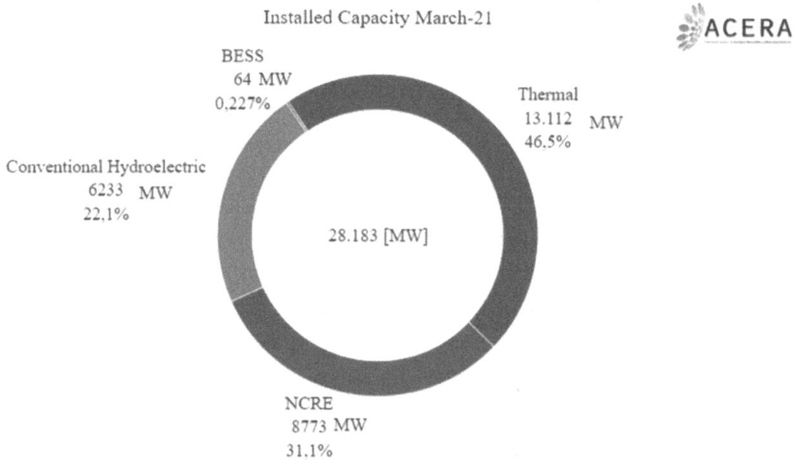

Fig. 4.1 Renewable energy capacity in Chile in March 2021[30]

there are 7,454 distributed plants across the country accounting for 78.3 MW.

In addition to the initiatives relating to diversification of the energy matrix and decarbonisation, it has been agreed with coal power generators that twenty coal-fired thermal power plants be decommissioned and closed by the year 2040. As of today, and according to the planning system data, five coal-fired thermal power plants have been shut down accounting for a total of 570 MW, all of them already replaced by NCRE.[31]

[30] ACERA, 2021. *Statistics Generation Sector Renewable Electrical Energy*, s.l.: s.n.

[31] Government of Chile, 2020. *Decree N°50*, Santiago: s.n.

4.4 Projections

4.4.1 Energy Needs Projections in the Mining Industry

According to the Chilean Copper Commission (COCHILCO) in its report 'Projection of electrical energy consumption in copper mining 2020–2031',[32] energy consumption in the mining industry will keep growing mainly due to the development of future projects, the enlargement of existing operations, and increasing production costs, among others. Energy consumption will increase from about 25 TWh in 2020 to approximately 33.4 TWh in 2031, which means that in the current generation capacity scenario, it will be necessary to add at least 8,400 GWh to the national electricity grid by 2030.

c. Projections on electricity generation capacity from NCRE sources.

Consistent with the numbers and studies presented by the Ministry of Energy, the country has the potential to generate around 1,800 GW from renewable technologies, about 70 times the current total installed electricity capacity of the country. This could be distributed throughout the territory, with great solar potential in the north of the country and wind power in the south.[33] This would consist of 509 GW (concentrated solar power), 1,180 GW (solar PV), 191 GW (onshore wind), and 6 GW (hydro). It is foreseen that power production costs will drop as generation capacity increases over the years, this is in addition to energy storage and transmission infrastructure, factors that will contribute to, and generate new economic opportunities for the country, such as the production and commercialisation of 'green hydrogen'[34] promoted by the Chilean government.[35]

---

[32] COCHILCO, 2020. *Gross Domestic Product by Economic Activity, at Current Prices*, Santiago: s.n.

[33] Energy Ministry, 2020. *National Green Hydrogen Strategy*, Santiago: Energy Ministry of Chile.

[34] Hydrogen produced with low-cost renewable energy.

[35] Energy Ministry, 2020. *National Green Hydrogen Strategy*, Santiago: Energy Ministry of Chile.

Following the deployment of renewable technology throughout the country, solar and wind power technologies are quickly maturing, and it is expected that by 2030 around 70% of total generated power will come from renewable sources.[36] Such achievements have led to further objectives and complementary applications being developed, such as using this new source of generated energy for alternative activities.

d. Foreseen transformations based on the development of NCRE power sources.

The rapid and thorough transformation witnessed in the Chilean energy sector was mainly prompted—together with regulatory changes—by Chile's geographical conditions and its vast potential for solar and wind power generation.[37] However, it should be understood that these transformations are only the first stage in series of required steps to come.

As Cristian Bustos stated,[38] this first stage stands out for transforming the energy framework and the increasingly widespread presence of large-scale renewable energies, with the subsequent aim of later exporting to neighbouring countries or complementing it with storage. The massification of renewable energies will probably extend to small-scale distribution systems that can then be inserted into the distribution networks.

In the second stage, end users could undergo a transformation due to continuous reductions in renewable generation costs.

The third transformation would be related to electromobility. Thus, the current downward trend in technology costs (see Fig. 4.2), together with the massive incentives observed worldwide for electromobility, suggest that this type of transportation could eventually be a feasible alternative in Chile due to developments in the first two transformations.

---

36 Ibid.

37 Ibid., 4.

38 Salinas Quintana, P., 2018. *The Environmental Path: Challenges and Projections for a Future Chile*, Santiago: s.n.

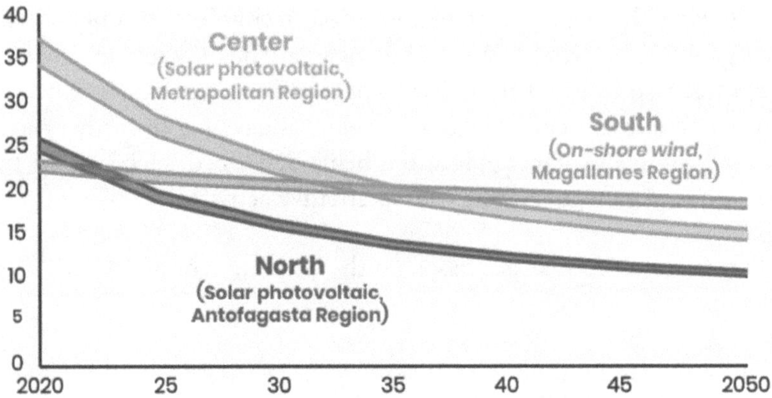

Fig. 4.2 Expected costs of renewable electricity (USD/MWh) in Chile[39] (Mckinsey and Company*)

4.4.2 Green Hydrogen

Just as the country builds a national strategy to achieve security of energy supply through NCRE (in the power sector), in 2020 Chile implemented the 'National Green Hydrogen Strategy'. This was aimed at unlocking the massive potential from producing clean and cheap hydrogen fuel and its by-products. This strategy foresees green hydrogen as a profitable industry and provides a solution for deep decarbonisation, driven by environmental trends, falling renewable costs, strategic government push, and industry coordination.[40]

The strategy stipulates an initial stage (2020–2025) to lay the foundations of the industry, including domestic usage replacing imported fuels with ammonia and grey hydrogen used in oil refineries. These actions seek to generate know-how, develop talent, deploy infrastructure, and attract financing. Through these actions, the country would be better positioned to access export markets. The practical implementation is geographically focused in the Magallanes Region,[41] since it has one of

[39] Energy Ministry, 2020. *National Green Hydrogen Strategy*, Santiago: Energy Ministry of Chile.

[40] Energy Ministry (n 35).

[41] Located in the extreme south of the national territory. It is located in the southwestern part of the continent from 48° 36′ to 56° 30′ south latitude and between meridians 66° 25′ and

the best wind resources nationwide, reaching an estimated potential of 126,000 MW. This could potentially be translated into large capacity renewable powered hydrogen production, reaching more than 10 million tons per year. Chile could go from the current 0.1% of global production to approximately 13%.[42]

The second stage (2025–2030) will extend local uses and start export activities. This stage is seen as essential to scale into global markets, ensuring industrial developments and creating tangible demand from green hydrogen and its derivatives. The third and last stage of the strategy will seek to establish long-term massive scale-up production of green hydrogen and promote decarbonisation at the international level. The focus is to expand as a global supplier, becoming leaders in exporting green hydrogen and derivatives.

In line with this long-term strategy, it is worth noticing that according to Mckinsey and Company, Chile has the potential to provide a significant share of the world's green hydrogen, focusing in areas such power generation, transportation, building heating and power, industry, new and existing feedstock uses.[43] They have proposed to fulfil those projections, taking into account the following next steps: articulating an overall vision; defining a stable regulatory framework; building multiparty coalitions, attracting funding, technology, international cooperation; and building industry support.[44]

4.4.3 Industrial Water Consumption

A resource that will become increasingly important in electricity consumption, particularly in the north of the country, is the use of seawater which, must first be desalinated and then sent to remote mining operations in the mountains. This relates specifically to the increase in the concentration process in mining operations, a highly intensive

75° 40′ west longitude: Library of Chilean National Congress: https://www.bcn.cl/siit/nuestropais/region12.

[42] Energy Ministry (n 35).

[43] Energy Ministry, 2020. *National Green Hydrogen Strategy*, Santiago: Energy Ministry of Chile.

[44] Energy Ministry (n 35).

process requiring substantial volumes of water, a particularly scarce resource in the desert regions of Antofagasta and Atacama where the large mining operations are located.

Indeed, in the study 'Forecast for water consumption in the copper mining industry, 2018–2029',[45] it was estimated that the consumption of seawater would practically double between 2018 and 2029, going from an estimated seawater consumption of 5.7 m3/sec to 10.9 m3/sec. This will add extra pressure to the electrical energy grid, bearing in mind that nowadays the energy-intensive mining sector represents more than 33% of total national electricity consumption.

4.4.4 Electromobility

According to the planning mechanism used by the Government of Chile, in 2017, the new 'National Electromobility Strategy' was presented. This document establishes specific goals for the use of electric vehicles. This strategy sets out that by the year 2040, 100% of urban public transport will be electric and by 2050, private electric vehicles will account for 40% of the automobile fleet. Therefore, it inevitably puts pressure on generation activities and electric installation capacity.

4.5 Conclusion

As expressed throughout this chapter, Chile's energy sector over the last 17 years could be described as very dynamic. This has been driven in large part by the country's search for security of energy supply, adjustment of the legal framework, seeking better infrastructure, incentivising diversification of the energy matrix, focusing on electricity market governance and the challenges of integrating renewable energies to the electricity mix. Furthermore, together with the introduction of energy regulation reforms, different governments since 2004 have proposed and implemented long-term energy policies covering areas of interest such as

[45] COCHILCO, 2018. *Forecast for water consumption in the copper mining Industry, 2018–2029*, Santiago: s.n.

energy efficiency, the promotion of NCRE sources, diversification of the electricity matrix, strengthening the design of the transmission system, and addressing various market challenges.

This long-term process was not carried out randomly. On the contrary, it required planning and permanent support from technical organisations, for example, the IEA and the International Renewable Energy Association (IRENA), among others.[46] The regulatory framework was progressively adjusted, always with the intention of attracting investment without the provision of subsidies. It is a process that has had many positive results so far and has been concerned with maintaining permanent contact and community participation.

A relevant point that should be kept in mind is that the reform process was favoured by the geographical characteristics of the country, with a predominance of solar photovoltaic and wind power technologies, which in turn facilitated the implementation of initiatives and investment in the area.

The current status and the results of the actions taken more than a decade ago can only now be viewed in terms of such quantifiable effects. Namely, that the incorporated technologies are capable of making a difference from the initial state of energy scarcity.

For example, today 20% of total national electricity is generated through NCRE and it is projected that by 2030 this will reach 70%. Put simply, over two-thirds of the country's energy matrix will be from renewable energy sources.

This process required effort and commitment, particularly from a succession of governments, since the benefits of a government's actions do not materialise until a number of years later.

Given the review of the current regulatory and practical scenario, it seems that in the future it will be necessary to adjust the existing regulations or incorporate new ones so as not to slow down the development of the initial stage promoted by the new regulatory framework.

One such example, is the National Electricity Coordinator that has come under questioning from power generation companies, as they claim

[46] IRENA, 2020. *Global Renewables Outlook: Energy transformation 2050*, Abu Dhabi: s.n.

the Coordinator's lack of understanding regarding the Annual Transmission Expansion Plan, could be putting at risk the security of energy supply due to slow decision-making, poor management and planning, plus a lack of transparency, since its independence seems to have become a barrier in technical accountability.

The regulatory, institutional, and technical development elements that have been discussed here, allow us to say that in general terms, the decisions made to date have complied with the initial premise of the search for energy security. In fact, today energy of domestic origin is available in a sufficient quantity to provide relief from a historically expensive, polluting, and third-party dependent national energy matrix.

Likewise, and considering the projections provided by both the authorities and investors, it is expected that NCRE installed capacity and generation output will continue to grow. With it, other initiatives only possible with long-lasting and cheap fuel, will hopefully progress as well, including automotive technological change, the desalination of seawater for industrial and human use, and the production of green hydrogen.

5

Venezuela: Petroleum Industry and Regulation

César R. Mata-García

5.1 Introduction[1]

During the last two decades, the Venezuelan Oil & Gas industry has suffered one of the most significant legal-technical-managerial consequences of its entire petroleum history: an excessive control and regulation by the State. The consequence of this excessive control is reflected in Venezuela's increasingly declining crude-oil production, which, among other reasons, is the immediate consequence of a lack of investment and opportune decision-making, which together have eroded the trust and credibility of investors in Venezuelan State institutions.

[1] The cut-off date for the update of materials for this chapter was on April 25, 2021.

C. R. Mata-García (✉)
Caracas, Venezuela
e-mail: cmatagarcia@gmail.com

G. Wood and J. F. Neira-Castro, *From Fossil Fuels to Low Carbon Energy Transition*, Energy, Climate and the Environment,
https://doi.org/10.1007/978-3-031-00299-1_5

In fact, evidence suggests that the above-indicated lack of investments has resulted in: (1) according to the Organization of the Petroleum Exporting Countries (OPEC), a decline, in less than two decades, of national crude-oil production from more than 3.1 MMBD to just above 480 MBD, and (2) passing from an incipient industry of natural and associated gas (i.e. for industrial, domestic, vehicular, thermo-generation purposes) to an almost non-existent gas sector. The petrochemical industry is also included within this scenario due to being currently inoperative and highly dependent on this gas as its main source of raw material. A high percentage of gas in Venezuela is unfortunately flared or vented, according to Petroguía (2019), in volumes greater than 2,000 MMSCFD.

Excessive State regulation (understood as the extemporary promulgation of 'unnecessary' and 'decontextualised' legal principles over determined economic activities to consolidate the State's control) derived, in the last 20 years, according to Di Martino (2019), from a 'dogmatised' and 'socialist' State administration to one that fully controlled its economy and industry.

The latter situation has been further aggravated by the 'lack of professionalism' needed to manage the petroleum industry by Venezuela's State institutions and the necessary officials required to guarantee the optimal performance of the said industry.[2] Within this context, it is apt to highlight the *Rule of Law Index of 2020*, elaborated by the World Justice Project (which establishes the Rule of Law as when all activities of a given country are performed in accordance with pre-established norms) that ranks Venezuela at 128 out of 128 with a score of 0.27 points, out of a maximum score of one.

The above-described situation has resulted, according to Santos (2017), in a minimum level of investment in the hydrocarbon industry and the impoverishment of the country. Simultaneously, this situation has caused integral failures in operational aspects such as: infrastructure and logistics, financial and corporate, contractual, protection of technology, management and business, labour force and Venezuelan diaspora,

[2] See for more detail: C. R. Mata-García (2018) *Legal Diagnosis on the Venezuelan Oil & Gas Industry*, https://www.linkedin.com/pulse/legal-diagnosis-current-venezuelan-oil-gas-industry-mata-garcia-phd/ (Last visit on 12/04/2021).

physical security and protection of the personnel and company facilities, promotion of national and international investments and a lack of international State presence before international bodies and forums, among others.

All the above aspects have affected the level of crude-oil production in Venezuela and have affected, in consequence and according to CEDICE Libertad (2021), among other things: the delivery of basic (public) services such as electricity, water, medicine, social security and food in a continuous and efficient manner by the State.

In this context, the present chapter seeks to reflect on how excessive control and regulation by the Venezuelan State has affected, in a general manner, the hydrocarbons industry, including the low level of crude-oil production, as well as the optimal and rational exploitation and use of the country's main raw material for export such as natural or upgraded hydrocarbons.

The following paragraphs identify and analyse how, through recent Venezuelan petroleum history (with particular emphasis on the last two decades), excessive control and regulation by the State has impacted the petroleum industry. Additionally, the manner in which the lack of decision-making has limited opportunities for recovery, stabilisation and development of said hydrocarbons industry in the country will be examined.

The foregoing exercise includes identifying collateral damage caused to Venezuela as a country that relies heavily on petrochemicals and, as a consequence, to a society that has become highly dependent on foreign currency. The present paper does not pretend to cover, in detail, all legal aspects related to its title, but it does intend, on the contrary, to ignite the readers' interest and encourage undertaking of complementary research and reading of specialised papers on this topic.

As a conscious decision, this chapter does not address, discuss or debate energy transition, climate change or the adaptability of the energy, oil or gas sectors in Venezuela, because these things have yet to happen. The country's political and economic problems keep us focused on the past and present but with a short-sightedness for future change.

5.2 Evolution of Hydrocarbon Regulation in Venezuela Until 1998: Summary of the Main Aspects

The regulatory history of the Venezuelan petroleum industry goes back formally to the *Hydrocarbons and other Mineral Fuels Law*[3] of 1943. This was in turn the result of a legislative and historical evolution that started on June 19, 1920, through the first hydrocarbons law entitled *Hydrocarbons and other Mineral Fuels Law*. According to Mata-García (2016), the law of 1943 indicated a meaningful change in the relationship between concessionaires and the State.

The Law of 1943 not only allowed more State participation in the petroleum business, but also a bigger government take through the creation by law of a petroleum royalty (e.g. a minimum of 16 2/3% over the net production) and the implementation of the now famous fiscal formula known as the 'fifty-fifty' (i.e. the promotion of an equal participation between the concessionaires and the State in the petroleum business). In 1943, crude-oil production in Venezuela was below 500 MBD, increasing to approximately more than 3.5 MMBD by 1969.

After more than 30 years, as a result of the waves of nationalisation that took place in many oil producing countries such as Indonesia (1965), Algeria (1971), Iraq and Libya (1973) and under the guidelines of the *New International Economic Order* discussed in the United Nations in the 60s and 70s, Venezuela passed a group of laws destined to take control, through increasing nationalisation, over the national economy's most productive and profitable activity: the petroleum industry.

Within this context, in 1971 what could be considered, according to Silva Calderón (2011), as the 'first step' of the Venezuelan petroleum industry's nationalisation was promulgated: the *Law about Goods Affected to Reversion in the Hydrocarbons Concessions*,[4] known locally as the 'Reversion Law'. This same year, the States control over the gas industry

[3] Ley de Hidrocarburos y Demás Minerales Combustibles. G.O. No. 32 del 13 de marzo de 1943.

[4] Ley sobre Bienes Afectos a Reversión en las Concesiones de Hidrocarburos. G.O. No. 59.577 del 06 de agosto de 1971.

materialised, through the *Law that Reserves to the State the Gas Natural Industry.*[5] During this period, crude-oil production declined from approximately 3.5 MMBD to 3.2 MMBD.

With these two laws, it only remained for the State to take control over the hydrocarbons (petroleum) sector. In this context, on August 29, 1975, the *Organic Law that Reserves to the State the Industry and Commerce of Hydrocarbons*[6] was promulgated, locally known as the 'Nationalisation Law'. Through this law, the State not only reserved for itself the right to exercise control over all activities related to the liquid hydrocarbons' chain of value, but also consolidated the State's control and control over the gaseous hydrocarbons industry started in 1971. Throughout the process of nationalisation, crude-oil production declined from approximately 3.2 MMBD in 1971 to 2.2 MMBD in 1995.

Nevertheless, despite the above-mentioned legal reserves in favour of the State over the exercise of all activities related to both liquid and gas hydrocarbons, the *Nationalisation Law* of 1975 established, by way of exception, the possibility of performing said activities through participation and support of the private sector (national and international).

The referred exception was in Article 5 of the aforementioned 'Nationalisation Law', coined as the 'Escape Clause'. Through the interpretation and application of this Article, the process of 'Petroleum Opening' took place in the 90s. This process was mainly materialised through three different contractual models: (1) operating services contracts, (2) strategic associations (joint ventures) and (3) production sharing agreements.

According to Mata-García (2016), the opening of the domestic market to refined products was also part of this process through the promulgation of the *Organic Law of Internal Market Opening of Fuels and Other Products for Automotive Vehicles*[7] of 1998. During the indicated opening

[5] Ley que Reserva al Estado la Industria del Gas Natural. G.O. No. 29.594 del 26 de agosto de 1971.

[6] Ley Orgánica que Reserva al Estado la Industria y el Comercio de los Hidrocarburos. G.O. No. (E) 1.796 del 29 de agosto de 1975.

[7] Ley Orgánica de Apertura del Mercado Interno de la Gasolina y Otros Combustibles derivados de los Hidrocarburos para el Uso en Vehículos Automotores. G.O. No. 36.537 del 11 de septiembre de 1998.

process, crude-oil production increased from approximately 1.8 MMBD in 1985 to 3.4 MMBD in 1998,.

After almost 15 years, the process of petroleum opening was subject to a revision by the National Executive. This decision together with the above-indicated regulatory and legislative initiatives has had a significant impact over the fluctuation of crude-oil production in Venezuela, as can be appreciated in Fig. 5.1.

As can be appreciated from the graph, fluctuations in crude-oil production has varied (positively and negatively) each time that there has been a legislative/regulatory change that directly affected the Venezuelan hydrocarbons sector. As was pointed out by Asdrúbal Oliveros (2019), it seems that '… [t]he foreign investment [in the hydrocarbons sector] is especially sensitive to [legal] expectations …' (Emphasis of the author).

Within this context, the last two decades have special relevance for the analysis for this papers topic. Thus they allow us to appreciate how levels of crude-oil have fluctuated in a representative manner during the last 20 years. In fact, there has been a special tendency towards a slowed-down and constant decline of the volumes of crude-oil produced in the country, shown by production figures of, 3.1 MMBD in 2000 to just above 520 MBD in 2021.

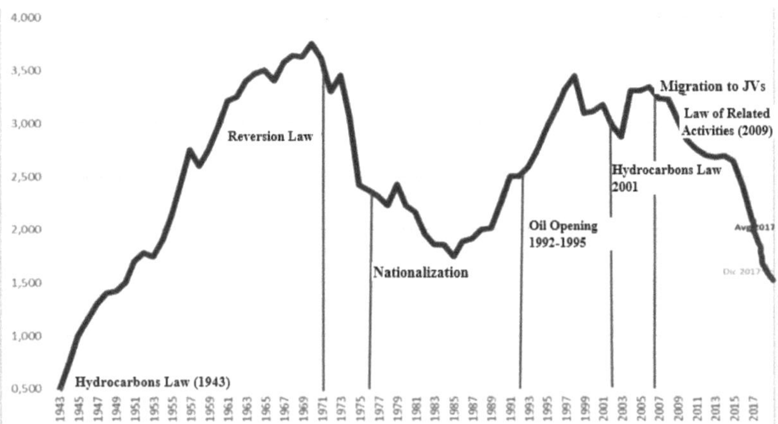

Fig. 5.1 Venezuela's Oil production (Petroleum Chamber of Venezuela, 2019)

Obviously, the causes or reasons for such fluctuation in Venezuelan crude-oil production are due, as has been mentioned throughout this paper, to the various political, economic, labour, managerial and legal factors, the 'latter factor' being this present paper's principal focus of analysis for reasons and limitation of space.

In legal terms, this suggests that said crude-oil declinature has been a direct consequence of excessive State regulation during the last two decades which translates into a notorious lack of legal certainty and stability for investors. For this reason, it is the author's responsibility to clarify that whenever legal certainty is mentioned, the promulgation (or the correct application) of a new or an existing norm or law is not the only aspect referred to. On the contrary, legal certainty also includes the human capacity to make good decisions. This contributes towards creating and maintaining transparent relationships between the State and the other players who are part of the Venezuelan hydrocarbons industry.

From all the above, it corresponds to the following paragraphs to highlight the main areas of legal evidence, giving a solid grounding to the argument about excessive regulation and control by the Venezuelan State and its impact on the petroleum industry during the last 20 years.

5.3 Excessive Regulatory Activity of the State Over the Hydrocarbons in Venezuela During the last Two Decades: A General Overview

Since 1999, after almost 60 years from the promulgation of the 'Hydrocarbons Law' of 1943 and after almost 25 years from the promulgation of the 'Nationalisation Law' of 1975, the hydrocarbons sector in Venezuela has been subjected to a significant proliferation of new regulatory dispositions, either by new laws or by reforms or amendments of existing laws.

Within this group exist two laws of utmost importance. These currently rule the said Venezuelan sector: the *Organic Hydrocarbons*

Law[8] of 2001 (hereinafter the 'Petroleum Law') and the *Organic Law of Gaseous Hydrocarbons*[9] of 1999 (hereinafter the 'Gas Law'). It is opportune to highlight that the levels of crude-oil production in Venezuela by 1999 were approximately 3.3 MMBD.

Additionally, in December of 1999 through popular referendum, a new constitution was approved in Venezuela entitled the 'Constitution of the Bolivarian Republic of Venezuela'.[10] Within the context of the hydrocarbons, the new constitutional text stipulated three important guiding principles: (1) the ownership of the Republic over its mineral and hydrocarbons reservoirs (Article 12); (2) the reservation of the petroleum industry to the State, through an organic law (the gas industry is not mentioned) (Article 302) and (3) the constitutional reserve of all shares (100%) of the national petroleum company known as PdVSA (only the headquarter) to the State, excluding all its affiliates (Article 303).

Within the framework of the above-mentioned regulations, there can also be found a 'proliferation' of enabling laws that were granted by the Legislative Power to the Executive Power during the last 20 years. Within these laws, there were the two enabling laws that gave legal ground to both the Gas Law of 1999 and the Petroleum Law of 2001, respectively.

On one hand, there was the *Organic Law that Authorises the President of the Republic to Pass Extraordinary Measures on Economic and Financial Matters Requested by the Public Interest*[11] dated to April 26, 1999, which was used to promulgate the Gas Law in the period between 1999 and 2000. On the other hand, there was the *Law that Authorises the President*

[8] Decreto No. 1.510 con Fuerza de Ley Orgánica de Hidrocarburos. G.O. No. 37.323 del 02 de noviembre de 2001.

[9] Decreto con Rango y Fuerza de Ley Orgánica de Hidrocarburos Gaseosos. G.O. No. 36.793 de fecha 23 de septiembre de 1999.

[10] Constitución de la República Bolivariana de Venezuela. G.O. No. (E) 36.860 del 30 de diciembre de 1999.

[11] Ley Orgánica que Autoriza al Presidente la República para Dictar Medidas Extraordinarias en Materia Económica y Financiera Requeridas por el Interés Público. G.O. No. 36.687 del 26 de abril de 1999.

of the Republic to pass Decrees with force of Law on the Topics therein Delegated,[12] dated on November 13, 2000, used to promulgate the Petroleum Law in the period between 2000 and 2001.

Both enabling laws 'exceptionally' and 'illimitably' empowered the National Executive to unify the hydrocarbons regulatory framework in Venezuela in just one text. Nonetheless, the referred empowerment was partially fulfilled due to this, and as indicated above, there still exit two laws in Venezuela to separately regulate the oil and gas activities.

Regarding the gaseous hydrocarbons, the new Gas Law was promulgated in September of 1999, which, differed from the previous Gas Law of 1971 (that reserved to the Venezuelan State the totality of the gas industry), and allowed private investment of up to 100% in all those activities that conformed the gas chain of value, e.g. from gas exploration and production until its final retail.

However, according to Mata-García (2018), one of the main problems in correctly implementing the above-mentioned new Gas Law has been the subsidised price of natural gas in the national market, i.e. while the price for millions of BTU is of ± US$ 2.92 abroad, in Venezuela it is approximately ± US$ 0.0010 (BsS. 3.46789). Despite the hyperinflation in Venezuela, the price of gas has not been updated by the Ministry of Petroleum since 2006. This latter situation has been one of the main economic disincentives for investors who would bring fresh capitals to the gas sector in Venezuela.

Besides the above-indicated scenario referencing the commercial value of gas in the local currency (bolivars), there is also the issue of Venezuelan currency devaluation. This initially goes back to 1999, evidenced from a simple comparison of its monetary value from 1999 to 2021. For example, in 1999 one American dollar cost 570 bolivars, but in 2021, the same dollar cost about 2,700,000 sovereign bolivars.

The best manner to understand the impact of said devaluation is by taking into consideration the national hyperinflation and the bolivar's devaluations of 2008 and 2018. In 2008, the bolivar was

[12] Ley que Autoriza al Presidente de la República para dictar Decretos con Fuerza de Ley en las Materias que se Delegan. G.O. No. 37.076 del 13 de noviembre de 2000.

called 'Strong Bolivar[13] (BsF)' and three zeros were eliminated, and in 2018 the bolivar was renamed as 'Sovereign Bolivar[14] (BsS)' and five zeros were eliminated. Thus, without these two bolivar devaluations, the value of one American dollar, by April 2021, would be approximately 270,000,000,000,000 of the former bolivar value of 1999.

Regarding the regulation of liquid hydrocarbons, the Petroleum Law was promulgated in November of 2001 and amended in 2006.[15] This differed from the 'Nationalisation Law' of 1975, where all hydrocarbons' activities were reserved to the Venezuelan State. However, private participation was allowed as follows: (i) up to 49% in partnership with the State to carry out exploration and production activities (reserving for the State the majority of shares and the control of decisions of said commercial society), (ii) up to 100% in refining and external commercialisation activities (except the trade of natural crude-oil that must be sold to a State company), (iii) up to 100% in industrialisation activities and (iv) up to 100% in internal commercialisation or market activities. It is opportune at this point to highlight that after entering into force the Petroleum Law of 2001, crude-oil production declined from approximately 3.2 MMBD in 2000 to 2.9 MMBD in 2001.

Among the 'first steps' at reforming the guiding principles pre-established by the Petroleum Law of 2001, can be found some changes to the original text introduced by the amendment of the 2006 Petroleum Law. These changes were: (i) the regulation of the associated gas extraction, (ii) the approval by the National Assembly for any modification posterior to the original terms and conditions applicable to the operating of mixed companies (joint ventures), (iii) the new legal regime applicable to the royalty, (iv) the amendment of special taxes ruled by said Law and (v) the obligation of mixed operating companies to sell their crude-oil production to a State company, i.e. to PdVSA. By then, after the initial implementation of the Petroleum Law of 2001, crude-oil production temporally increased from approximately 2.9 MMBD to 3.0 MMBD.

[13] *Bolívar Fuerte.*

[14] *Bolívar Soberano.*

[15] Ley de Reforma Parcial del Decreto No. 1.510 con Fuerza de Ley Orgánica de Hidrocarburos. G.O. No. 38.493 del 4 de agosto de 2006.

Besides the above-mentioned 'steps' regarding new regulation on exploration and exploitation activities in Venezuela, on April 12, 2005, the National Executive, through the Ministry of Petroleum, ordered the PdVSA and the private sector to begin a negotiation process. This was known as 'the migration' where, former contractual schemes subscribed to during the Petroleum Opening in the 90s transitioned to the new business scheme of operating mixed companies stipulated by the above-indicated Petroleum Law of 2001.

The process of 'migration' took the then-existing schemes of business subscribed to during the term of the 'Nationalisation Law' of 1975 (e.g. the operating services agreements, the strategic associations (joint ventures) and the production sharing contracts), to the new business scheme of operating mixed companies established by the Petroleum Law of 2001 (i.e. a scheme where the State reserves for itself more than 50% of company's shares and control of decisions).

It is opportune to point out at this stage that the 'migration' process did not receive the full consent of all international petroleum corporations. Companies such as ExxonMobil and ConocoPhillips, considered the process inconvenient particularly in terms of the negotiation proposed by State representatives which were viewed as giving too much space for international claims including arbitrations. During the 'migration' process, crude-oil production declined from approximately 3.2 MMBD in 2006 to 2.8 MMBD in 2008.

Due to the notoriously political inherence of the Executive Power over the Legislative Power during the last twenty years (in particular between 2006 and 2007) and after the infructuous process of State negotiations with some private petroleum companies on existing contractual schemes, the National Executive decided to influence the Legislative Power to pass two new laws which focused on in the termination of those contracts subscribed during the process of Petroleum Opening. These two new laws were:

1. *Law to Regularise the Private Participation in the Primary Activities ruled by the Decree No 1,510 with Force of Organic Law of*

Hydrocarbons (2006)[16] whose main objective was to 'regularise' the participation of the private sector in exploration and exploitation activities for having violated the superior interests of the State and its sovereignty, and
2. *Law of Migration to Mixed Companies of the Strategic Associations of the Orinoco Petroleum Belt, as well as the Production Sharing Agreements* (2007)[17] the main objective of which was to 'force' the existing joint ventures to migrate to the new business schemes ruled by the Petroleum Law of 2001 (i.e. operating mixed companies in association with the State).

In 2007, these latter legislative initiatives consolidated in Venezuela the idea of 'renationalising' the hydrocarbons industry as a mechanism to increase the State's control over almost all the country's exploration and exploitation activities. During the process of these regulatory initiatives, crude-oil production jumped from approximately 2.8 MMBD in 2008 to 2.4 MMBD in 2015.

The increasing State control over its hydrocarbons industry also escalated to other petroleum areas related to exploration and exploitation such as the operating services, external commercialisation, petrochemical and other areas. For example, this control continued to consolidate the idea of 'statism' through another group of laws:

1. The *Organic Law that Reserves to the State the Related Activities to the Primary Activities of Hydrocarbons* (2009),[18] which eliminated, by reserving it to the State, the possibility of contracting private investors directly to carry out activities such as injection of water, vapour or gas, compression of gas and all those activities related to actions in the Maracaibo Lake,

[16] Ley de Regularización de la Participación Privada en las Actividades Primarias Previstas en el Decreto N° 1.510 con Fuerza de Ley Orgánica de Hidrocarburos. G.O. N° 38.419 del 18 de abril de 2006.

[17] Ley de Migración a Empresas Mixtas de los Convenios de Asociación de la Faja Petrolífera del Orinoco; así como de los Convenios de Exploración a Riesgo y Ganancias Compartidas. Publicada en la G.O. N° 38.419 del 18 de abril de 2006.

[18] Ley Orgánica que Reserva al Estado las Actividades Conexas a las actividades primarias de Hidrocarburos del 2009. G.O. 39.173 del 07 de mayo de 2009.

2. the *Presidential Decree No 1,648 dated January 15, 2002*[19] that reserved to the State the international trade of hydrocarbons products,
3. the *Organic Law for the Development of Petrochemical Activities* (2015)[20] that reserved to the State the basic and intermediate petrochemical activities, and
4. the *Organic Law to Reorganise the Internal Market of Liquid Fuels* (2008)[21] that reserved to State the exercise of those activities related to the intermediation of liquid fuels supply by ground and aquatic transportation and by coasting.

The idea of developing the 'renationalisation' of the hydrocarbons sector in Venezuela was also extended to and completed with other fiscal laws and tributary characters such as:

1. the *Law of Special Economic Assignations derived from Mining and Hydrocarbons* (2010),[22] which established a special economic assignations regime derived from mining and hydrocarbon activities in benefit of the territorial political entities and the organisations of popular basis, in accordance with Article 156:16 of the Venezuelan Constitution,
2. the *Law that creates the Special Contribution for Extraordinary and Exorbitant Prices in the International Market of Hydrocarbons* (2013),[23] which implemented a set of measures to prevent the fluctuation of oil-income that may affect the fiscal and monetary equilibrium of the country, and

---

[19] Decreto Presidencial Número 1.648 del 15 de enero de 2002. G.O. No. 37.429 del 24 de abril de 2002.

[20] Ley Orgánica para el Desarrollo de las Actividades Petroquímicas del 2015. G.O. No. (E) 2.171 del 30 de diciembre de 2015.

[21] Ley Orgánica de Reordenamiento del Mercado Interno de los Combustible Líquidos de 2008. G.O. No. 39.019 del 18 de septiembre de 2008.

[22] Ley de Asignaciones Económicas Especiales Derivadas de Minas e Hidrocarburos. G.O. Nº 5.991 Extraordinario del 29 de julio de 2010.

[23] Ley que crea Contribución Especial por Precios Extraordinarios y Precios exorbitantes en el Mercado Internacional de Hidrocarburos. Publicada en la G.O. Nº 40.114 del 20 de febrero de 2013.

3. the *Organic Law Related to the National Savings Fund of the Working Class and the Popular Savings Fund* (2012),[24] for which the main objective was to establish the 'populist' basements to create alternative mechanisms for payment of the social debt with the public sector workers and the promotion of savings.

Additional to the latter set of laws, the Venezuelan National Executive, within its policy of expanding control over the hydrocarbons sector, also promulgated some instruments of sub-legal rank in the same sense. Some of these instruments were:

1. *Resolution Number 005/2014*[25] from the Office of the General Attorney dated February 4, 2014, where through the argument of 'reinforcing'[26] the Public Procurement Law, it was agreed to 'modify' said Law by deciding: (a) to diminish the due amount for the inscription in the National Register of Contractors, (b) to request the inscription of interested people to participate in open international tenders, (c) to diminish the due amount to elaborate reports for the process of consulting prices and (d) to diminish the due amount requested for the concept of social responsibility.
2. *Resolution Number 019*[27] from the Ministry of Internal Affairs, Justice and Peace dated January 13, 2014, through which was requested a pre-authorisation from the Ministry of Petroleum to incorporate 'any' commercial society whose main social objective is related to hydrocarbons activities.

[24] Ley Orgánica Relativa al Fondo de Ahorro Nacional de la Clase Obrera y al Fondo de Ahorro Popular: G.O. N° 39.915 del 04 de mayo de 2012.

[25] Resolución No. 005/2014 de la Procuraduría General de la República 04 de febrero de 2014. G.O. No. 40.349 del 05 de febrero de 2014.

[26] The question here is how can be justified that an entity of the State freely decides to change a formal law without having faculties to do so nor following the pre-established rules designed for this purpose. Is it not illegal or unlawful?

[27] Resolución No. 019 del Ministerio del Poder Popular para Relaciones Interiores, Justicia y Paz del 13 de enero de 2014. G.O.40.332 del 13 de enero de 2014.

3. *Resolution Number 164*[28] from the Ministry of Petroleum dated December 6, 2017, through which the National Executive established a '*Regime to Review and Validate* all National and International Contracts subscribed by PdVSA, its affiliates and mixed companies where PdVSA has shares' (Emphasis added).

Notwithstanding the foregoing regulatory initiatives, not only did the Executive and Legislative Powers directly influence the regulation over hydrocarbon activities in Venezuela, but the Judicial Powers also did and so, through its maximum authority such as the Supreme Tribunal of Justice. In this case, the Constitutional Chamber of said Tribunal, through the Decision Number 156[29] dated March 29, 2017, decided that the constitutional competence of the Venezuelan National Assembly to approve contracts of 'National Interests' had passed to the said Chamber. The Decision can be summarised as follows:

> ... 4.4.- It is warned that while it persists the situation of disrespect and invalidity of the National Assembly actions, *this Constitutional Chamber guarantees that the parliamentary competences are directly exercised by this Chamber or by the entity that it appoints, to ensure the Rule of Law. ...* (Emphasis added)

It should be highlighted that one of the main objectives of this Decision was to authorise the Supreme Tribunal of Justice, in replacement of the natural competence of the National Assembly, to approve terms and conditions applicable to the creation and performance of the operating mixed companies (joint ventures) as was the case of Petrosur, S.A.[30]

The above-mentioned Decision Number 156 derived from a 'Resource of Interpretation' that was presented before the Supreme Tribunal of Justice by a PdVSA's affiliate: the Venezuelan Corporation of Petroleum,

[28] Resolución 164 del 06 de diciembre de 2017 del Ministerio del Poder Popular de Petróleo. G.O. No. 41.294 del 06 de diciembre de 2017.

[29] Sentencia No 156 del 29 de marzo de 2017, Sala Constitucional del Tribunal Supremo de Justicia. Asunto: Recurso de interpretación. Solicitante: Corporación Venezolana de Petroleo, S.A. (CVP).

[30] Petrosur. S.A. is an operating mixed company (a joint venture) between CVP (a PdVSA's affiliate) and the company *Stichting Administratiekantoor* Inversiones Petroleras Iberoamericanas.

S.A. (CVP). The argument was a conflict between the Article 187:24 of the Venezuelan Constitution and the Article 33 of the Petroleum Law of 2001. The parliamentary competences were restored to the Venezuelan National Assembly after a clarification of said decision issued on April 1, 2017. However, within the context of these legal events, crude-oil production in Venezuela declined from approximately 2.5 MMBD in 2015 to 1.5 MMBD in 2017.

Now, due to the chronic socio-economic situation that Venezuela has faced during the last 20 years, which has also generated a severe institutional and humanitarian crisis, many Western countries concerned about the above-indicated situation in Venezuela decided to take international, political and economic actions to help Venezuela to restore the Rule of Law and, therefore, its democracy.

As part of the above-mentioned international actions, the USA's administration has granted, since 2015, seven Executive Orders[31] to sanction the Venezuelan State and Government, including the Central Bank of Venezuela and PdVSA. One of the main objectives of said Orders was to block all assets of PdVSA in USA, as well as to limit any American persons or entities carrying out business with PdVSA, its affiliates and any other company that possessed more than 50% of the social capital or interests, including its affiliate in USA: CITGO.[32]

Before the wide scope of the referred sanctions,[33] the American administration has also granted more than 35 general licences to limit and order American entities and third parties to close down their financial operations and activities with the PdVSA and its affiliates. By way of example, it is worthy to mention: General Licence Number 7 that prohibited any kind of exportation of PdVSA and its affiliates to and

[31] The Executive Orders are: (1) Executive Order No. 13.692, March 8, 2015, (2) Executive Order No. 13.808, August 24, 2017, (3) Executive Order No. 13.827, March 19, 2018, (4) Executive Order No. 13.835, May 21, 2018, (5) Executive Order No. 13.850, November 1, 2018, (6) Executive Order No. 13.857, January 25, 2019, (7) Executive Order No. 13.884, August 5, 2019.

[32] Citgo Petroleum Corporation also known as CITGO is an affiliate of PdVSA in USA.

[33] See for more information: U.S. Department of the Treasury, *Venezuela-related Sanctions*, https://www.treasury.gov/resource-center/sanctions/Programs/pages/venezuela.aspx (Last visit on April 22, 2021).

from the USA, General Licence Number 8 that prohibited the exportation of diluents from USA to Venezuela, General Licence Number 9 that prohibited the commercialisation of PdVSA's bonds among American citizens, General Licence Number 10 that prohibited to resell or export refined products from Venezuela to the USA, among others. It is important to highlight that at the moment of implementing the indicated sanctions, crude-oil production was approximately 1.1 MMBD.

It should also be pointed out that, besides the above-mentioned legal elements, the Venezuelan hydrocarbons industry has been severely affected by general electricity failures that were provoked due a lack of maintenance and investment in the national grid. One of these failures was reported by the Venezuelan newspaper El Nacional in 2019 and, according to non-official numbers, during this blackout crude-oil production decreased for 2019 below 480 MBD approximately.

On the other hand, within the ongoing institutional conflicts in Venezuela and before the presidential lack of power argument levelled at the republic after the election of May 2018, the president of the National Assembly temporarily assumed the presidency of the country with the objective of recovering the Venezuela's institutionality. In this context, many plans were presented to the people by the interim government for the recovery of said institutionality as well as the Rule of Law.

Within the plans to recover the Rule of Law and democracy in Venezuela, in February 2019 the project entitled 'Plan País' was presented to the people by the interim government. This plan sought to reactivate the country's productive apparatus as well as take action to repair the collateral damage caused during the last 20 years, most notably, for the excessive control and regulation by the Venezuelan State over its productive sector, including the hydrocarbons sector.

In the framework of 'Plan País', a new hydrocarbons law was proposed (though yet not approved), of which the main objectives were: (1) to make the existing contract model more flexible to attract private investments, (2) to review the fiscal regime, (3) to create a Venezuelan Agency of Hydrocarbons, (4) to promote guaranties from private investors, (5) to prompt the settlement of petroleum disputes from national courts, (6) to guarantee a competitive salary in comparison with the salary of other petroleum companies abroad and (7) to create a patrimonial fund with petroleum incomes.

The end point of said (petroleum) plan was summarised, to among others: (1) guaranteeing the energy sovereignty and security for all Venezuelans, (2) promoting new technologies, and (3) to help transit to other sources of energy such as wind, solar, wave power. At the moment of presenting the referred 'Plan País' to the Venezuelan people, crude-oil production was below one million barrels per day.

In summary, it can be said that these regulatory experiences could serve as a lesson not only for Venezuela, but also to those countries where politicians do not act as 'public servants' but as 'dogmatic and circumstantial regulators' of socio-economic situations, including social crisis in detriment to national economic welfare and collective interest.

In this context, the following sub-section contains a table elaborated to illustrate to the reader the main historical moments of Venezuela's hydrocarbons sector and help to contextualise and understand the 'sui generis' character of the country's petroleum industry and regulation.

The idea of this exercise is to provide the reader with some guidelines (or warnings) to reflect on what has happened, what is happening and what could happen within one of the sectors that is considered most profitable for a country, the hydrocarbons industry. Furthermore, where the ideas of 'rentism' and 'statism' seem to prevail, from time to time, over the collective interest of a country.

5.4 General Comments on the Venezuelan Petroleum History and the Excessive Regulation by the Venezuelan State Over Its Hydrocarbons Industry

Venezuela is an atypical case in matters related to hydrocarbons regulations.[34] The country passed from being a worldwide reference on 'how to manage the petroleum industry' to a concerning reference about 'how

[34] The Venezuelan mining sector is under the same situation. See for more information: M. C. Acosta-García, (2020) Venezuela: Mineral Policy. In: Tiess G., Majumder T., Cameron P.

Table 5.1 Main features of key institutional changes in Venezuela's petroleum sector

1943 'Petroleum Reform'	1975 'Nationalization'	1990's 'Oil Opening'	2000's 'Plenty Petroleum Sovereignty'	2010–2015 Hidden opening/new financing scheme?	2019–2020 Proposal of a 'New Oil Regulation'
More participation of the State within the petroleum activities. Increase the Government Take	Control of all petroleum activities on part of the State	Participation of third parties within the Nationalized Oil Industry	Control of the State-Permanent Sovereignty over Natural Resources (UN Resolution 1803)	Need for deregulation and investment/new financing schemes with the State's allies	Need for private investments to integrally develop (and/or recover) the Oil and Gas Sector in Venezuela

a country should never manage', said industry.[35] To facilitate the reader's understanding, the author has summarised the regulatory history of Venezuela's petroleum sector in Table 5.1.

Table 5.1 summarises, in a general manner, the main historical moments regarding the most relevant regulatory aspects performed on Venezuela's hydrocarbons industry. It is important to highlight that said historical moments do not only include regulatory matters, but also the ineludible technical and operational consequences caused to the Venezuela hydrocarbons industry immediately resulting from excessive State control and regulation.

Additionally, it is necessary to take into consideration the lack of opportunities and effective decision-taking by Venezuelan institutions that have affected the day-to-day dynamic required by the operational activities of this kind of industry.

As pointed out by Stevens (2008), the successor frequently wants to undo previous government action, in particular when the exploitation

(eds) Encyclopedia of Mineral and Energy Policy. Springer, Berlin, Heidelberg, https://doi.org/10.1007/978-3-642-40871-7_247-1 (Last visit 27/04/2021).

[35] See for more details: A. Cheatham and R. Cara Labrador (2021) *Venezuela: The Rise and Fall of a Petrostate*, https://www.cfr.org/backgrounder/venezuela-crisis (Last visit 12/04/2021).

of natural resources is compromised. In fact, when there are recurring processes of regulation by a determined State to take control of its market or industry, the immediate effect is the inevitable deregulation of said sector in regards allowing in new and fresh capital.

Within this context, evidence of the effects provoked by constant regulatory changes is found in the (subjective) sensation of economic and juridical stability that private investors may feel with the host-state. For example, Venezuela is ranked 177 out of 178 as the most restrictive economy of the world, according to the *Economic Freedom Index of 2021*, elaborated by Heritage Foundation.

Nonetheless, according to Mata-García (2018), this historical moment that the Venezuelan hydrocarbons industry is living through must serve as a lesson for all petroleum sector players in the country. Furthermore, it should be seen as a lesson helping the shift to a new level that implies not only active and decentralised participation of the public sector, but also constant and consistent participation of the (national and international) private sector.

As a corollary to the above-mentioned experience of Venezuela's petroleum industry, it is fitting to warn that the problems in the country's current hydrocarbons industry, caused by political, managerial and legal factors are not the only unacceptable situation. Additionally, there is a need for being 'cautious' and 'foresighted' before, 'dogmatic' or enacting capricious changes of public policies that may sensibly affect the hydrocarbons industry as seems to have been seen recently with Mexico and its petroleum industry.

For example, Mexico (after nationalising its hydrocarbons industry in 1938 and opening it to private investors in 2013) has been recently swarming, according to Acosta García (2019), with the idea of '[t]he petroleum expropriation and rescuing the nation…'. In fact, according to Morgan Lewis (2021), Mexico is seeking to pass a new hydrocarbons law to control foreign investments in the petroleum sector.

This latter reflection invites us to consider from time to time, that the experience of one petroleum producing country in Latin American can be used as a reference for other petroleum producing countries in the region or even in the world and where ideas of 'rentism' and of 'statism' are still present in the mind of many politicians and civil servants.

5.5 Final Reflections

The experience of Venezuelan crude-oil production fluctuating from 3.1 MMBD to just above 480 MBD, in less than two decades, is, from an academic and practical point of view, a topic that requires more attention and analysis. Many experts agree on the sensibilities that lead to legislative and regulatory changes over investments in general.

The excessive regulation (understood as the extemporary and/or unnecessary promulgation of norms) plus the excessive control by the State on its productive apparatus, affect the development of the hydrocarbons industry as has been the case of Venezuela. The regulation of the socio-economic equilibrium should not jump from an 'inofficious' excessive regulation to 'unplanned' deregulation that could even promote cases of self-regulation.

The Venezuelan experience shows that the temptation of excessive regulating to control the market (or even to solve social crisis), should be avoided, in particular, over such economic activities as hydrocarbons whose financial nature requires periods of 10 or 20 years to fully recover the investment. Therefore, the author would like, through this article, to invite all players of the hydrocarbons industry to reflect on the points mentioned above, so the history and the learned experiences could bring significant ideas on how to design public policies, regulations and projects for the future.

Many of the consequences that affect the Venezuelan hydrocarbons sector are down to a lack of opportune and right managerial decision-making by State institutions, as well as on cultural and labour aspects that are derived from the lack of clear and consistent petroleum policies that have also generated legal and contractual problems. In fact, many of the technical and operational problems caused to the Venezuelan petroleum business should not only be corrected by drafting simple and straightforward laws (avoiding complicated regulations), but also through professional, appropriate, right and responsible decision-making.

The technical and regulatory problems of the Venezuelan petroleum industry are based on a question that seems to require more political goodwill than anything else, because this political action must, in consequence, bring a legal certainty and stability to the people. In this context, the legal certainty comes not only from the law, but also from the credibility and trust that particulars may have in the State's institutions.

In summary, it is urgent to review all those dogmatic and legal subterfuges created to divert the original principals of Venezuela's hydrocarbons laws, as well as to clarify in advance which petroleum policies to adopt in order to enact necessary future regulations. It is clear that laws obey public policies, so they must bring the legal certainty requested by all actors of the hydrocarbons sector, allowing them to elaborate with certainty their medium and long-term investments projects.

As mentioned before, Venezuela and its institutions have focused on the development of a very profitable sector that has provided substantial resources for the development of the country. However, despite a changing world, particularly relating to climate, the energy transition is not reflected in the discussions and development of Venezuela's energy sector.

References

2021 Index of Economic Freedom, The Heritage Foundation. https://www.heritage.org/index/ranking (Last visit on 09/03/2021).

Acosta García, I., *La Expropiación Petrolera y el Rescate de la Nación* (México, 2019). http://www.cambiodemichoacan.com.mx/columna.php?tipo=nc&id=51758 (Last Accessed 27/03/2021).

Acosta-García, M. C. (2020) *Venezuela: Mineral Policy*. In: Tiess G., Majumder T., Cameron P. (eds) Encyclopedia of Mineral and Energy Policy. Springer, Berlin, Heidelberg. https://doi.org/10.1007/978-3-642-40871-7_247-1 (Last Accessed 27/04/2021).

CEDICE Libertad (Centro de Divulgación del Conocimiento Económico para la Libertad). (2021) *Fallas en los servicios básicos, la calamidad que padecen los venezolanos* in todosahora.com. https://www.todosahora.com/venezuela/fallas-en-los-servicios-basicos-la-calamidad-que-padecen-los-venezolanos/ (Last Accessed 03/04/2021)

Cheatham, A. and Cara Labrador, R. (2021) *Venezuela: The Rise and Fall of a Petrostate*. https://www.cfr.org/backgrounder/venezuela-crisis (Last Accessed 12/04/2021).
Constitución de la República Bolivariana de Venezuela. G.O. No. (E) 36.860 del 30 de diciembre de 1999.
Decreto con Rango y Fuerza de Ley Orgánica de Hidrocarburos Gaseosos. G.O. No. 36.793 de fecha 23 de septiembre de 1999.
Decreto No. 1.510 con Fuerza de Ley Orgánica de Hidrocarburos. G.O. No. 37.323 del 02 de noviembre de 2001.
Decreto Presidencial Número 1.648 del 15 de enero de 2002. G.O. No. 37.429 del 24 de abril de 2002.
Di Martino, D. (2019) *How Socialism Destroyed Venezuela*. https://economics21.org/how-socialism-destroyed-venezuela (Last Accessed 23/03/2021).
El Nacional, *Se registra falla eléctrica en varios estados del país* (Marzo 2019). http://www.el-nacional.com/noticias/servicios/registra-falla-electrica-varios-estados-del-pais_273793 (Last Accessed 10/03/2021).
Ley de Asignaciones Económicas Especiales Derivadas de Minas e Hidrocarburos. G.O. Nº 5.991 Extraordinario del 29 de julio de 2010.
Ley de Hidrocarburos y Demás Minerales Combustibles. G.O. No. 32 del 13 de marzo de 1943.
Ley de Migración a Empresas Mixtas de los Convenios de Asociación de la Faja Petrolífera del Orinoco; así como de los Convenios de Exploración a Riesgo y Ganancias Compartidas. Publicada en la G.O.Nº 38.419 del 18 de abril de 2006.
Ley de Migración a Empresas Mixtas de los Convenios de Asociación de la Faja Petrolífera del Orinoco; así como de los Convenios de Exploración a Riesgo y Ganancias Compartidas. G.O. N° 38.632, de fecha 26 de febrero de 2007.
Ley de Reforma Parcial del Decreto No. 1.510 con Fuerza de Ley Orgánica de Hidrocarburos. G.O. No. 38.493 del 4 de agosto de 2006.
Ley de Regularización de la Participación Privada en las Actividades Primarias Previstas en el Decreto N° 1.510 con Fuerza de Ley Orgánica de Hidrocarburos. G.O. Nº 38.419 del 18 de abril de 2006.
Ley Orgánica de Apertura del Mercado Interno de la Gasolina y Otros Combustibles derivados de los Hidrocarburos para el Uso en Vehículos Automotores. G.O. No. 36.537 del 11 de septiembre de 1998.
Ley Orgánica de Reordenamiento del Mercado Interno de los Combustible Líquidos de 2008. G.O. No. 39.019 del 18 de septiembre de 2008.

Ley Orgánica para el Desarrollo de las Actividades Petroquímicas del 2015. G.O. No. (E) 2.171 del 30 de diciembre de 2015.

Ley Orgánica que Autoriza al Presidente la República para Dictar Medidas Extraordinarias en Materia Económica y Financiera Requeridas por el Interés Público. G.O. No. 36.687 del 26 de abril de 1999.

Ley Orgánica que Reserva al Estado la Industria y el Comercio de los Hidrocarburos. G.O. No. (E) 1.796 del 29 de agosto de 1975.

Ley Orgánica que Reserva al Estado las Actividades Conexas a las actividades primarias de Hidrocarburos del 2009. G.O. 39.173 del 07 de mayo de 2009.

Ley Orgánica Relativa al Fondo de Ahorro Nacional de la Clase Obrera y al Fondo de Ahorro Popular: G.O. N° 39.915 del 04 de mayo de 2012.

Ley que Autoriza al Presidente de la República para dictar Decretos con Fuerza de Ley en las Materias que se Delegan. G.O. No. 37.076 del 13 de noviembre de 2000.

Ley que crea Contribución Especial por Precios Extraordinarios y Precios exorbitantes en el Mercado Internacional de Hidrocarburos. Publicada en la G.O. Nº 40.114 del 20 de febrero de 2013.

Ley que Reserva al Estado la Industria del Gas Natural. G.O. No. 29.594 del 26 de agosto de 1971.

Ley sobre Bienes Afectos a Reversión en las Concesiones de Hidrocarburos. G.O. No. 59.577 del 06 de agosto de 1971.

Mata García, C. R. International Investment and Regulatory Disputes - Applying Principles of Administrative Law to Investor State Treaty Arbitrations (Editorial Scholars' Press, Germany, 2016).

Mata García, C. R., *Legal Diagnosis on the Venezuelan Oil & Gas Industry* (Caracas, 2018). https://www.linkedin.com/pulse/legal-diagnosis-current-venezuelan-oil-gas-industry-mata-garcia-phd/ (Last Accessed 12/04/2021).

Mata García, C. R., Manual de Derecho de los Hidrocarburos en Venezuela (Editorial Arte, Caracas, 2016).

Monedas de Venezuela, *Cronología: Devaluación del Bolívar, desde J.V Gómez hasta N. Maduro* (Junio 2019). http://www.monedasdevenezuela.com/articulos/cronologia-de-la-devaluacion-del-bolivar-venezolano/ (Last Accessed 02/04/2021).

Morgan Lewis, *Mexico Targets Foreign Investors in Oil and Gas Sector with Proposed Changes to Hydrocarbon Law* (Abril 2021). https://www.jdsupra.com/legalnews/mexico-targets-foreign-investors-in-oil-3200170/ (Last Accessed 11/04/2021).

Oliveros P., A. R., *Transición en Venezuela – Mito vs Realidad* (Caracas, 2019). https://www.linkedin.com/in/asdrubal-r-oliveros-p-797a1411/detail/recent-activity/ (Last Accessed 26/04/2021).

Organización de Países Exportadores de Petróleo (OPEP), *Boletines de Producción* (Viena, 2021). https://www.opec.org/opec_web/en/76.htm (Last Accessed 22/04/2021).

Petroguía, *Quema de gas" en Venezuela alcanza récord de 2.000 millones de pies cúbicos al día*, (Abril 2019). http://www.petroguia.com/pet/noticias/gas-natural/%E2%80%9Cquema-de-gas%E2%80%9D-en-venezuela-alcanza-r%C3%A9cord-de-2000-millones-de-pies-c%C3%BAbicos-al (Last Accessed 11/04/2021).

Resolución 164 del 06 de diciembre de 2017 del Ministerio del Poder Popular de Petróleo. G.O. No. 41.294 del 06 de diciembre de 2017.

Resolución No. 005/2014 de la Procuraduría General de la República 04 de febrero de 2014. G.O. No. 40.349 del 05 de febrero de 2014.

Resolución No. 019 del Ministerio del Poder Popular para Relaciones Interiores, Justicia y Paz del 13 de enero de 2014. G.O.40.332 del 13 de enero de 2014.

Rule of Law Index 2020, World Justice Project. https://worldjusticeproject.org/sites/default/files/documents/WJP-ROLI-2020-Online_0.pdf (Last Accessed 30/03/2021).

Santos, M. A. (2017), La inversión privada y el empobrecimiento sostenido de Venezuela,. https://www.ucab.edu.ve/wp-content/uploads/sites/2/2017/09/INV-IIES-REV-087-La-inversion-privada-y-el-empobreciminto-sostenido-de-Venezuela.pdf (Last Accessed 25/03/2021).

Sentencia No 156 del 29 de marzo de 2017, Sala Constitucional del Tribunal Supremo de Justicia. Asunto: Recurso de interpretación. Solicitante: Corporación Venezolana de Petroleó, S.A. (CVP).

Silva Calderón, A., La Reversión y la Nacionalización en el sistema legal de los hidrocarburos en Venezuela (Colección Sendero de Luz, Cátedra Petrolera "Dr. Gumersindo Torres" de la Universidad del Zulia, 2011).

Steven, P., *Oil Wars: Resource Nationalism and the Middle East,* CEPMLP 30 Anniversary Seminars, (June 2008).

U.S. Department of the Treasury, *Venezuela-related Sanctions.* https://www.treasury.gov/resource-center/sanctions/Programs/pages/venezuela.aspx (Last Accessed April 22, 2021).

6

The Mexican Energy Transition

Ernesto Beltrán Nishizaki

6.1 Introduction

The Mexican energy sector has been historically associated with nationalist sentiment and ideology. Since expropriation of the oil and gas industry in the latest 1930s, this sector has been understood as a property of all Mexicans from which national development depends. Therefore, to achieve this understanding, the State took action, partly to avoid national sovereignty and energy sovereignty being affected or made vulnerable.[1]

[1] The energy sovereignty concept based on the current administration's ideological policy should be understood as States Companies developing the oil and gas industry, with private participation viewed as a violation of state sovereignty.

E. B. Nishizaki (✉)
Galtec Asesores, Mexico City, Mexico
e-mail: ebeltrann@galtec.info

G. Wood and J. F. Neira-Castro, *From Fossil Fuels to Low Carbon Energy Transition*, Energy, Climate and the Environment,
https://doi.org/10.1007/978-3-031-00299-1_6

The oil and gas industry has been crucial for the State's revenue and, consequently, the Federal Annual Budget. Even today, the Ministry of Finance estimated for 2021 that the income from the oil and gas sector will likely account for 16.7% of the State's annual income,[2] equivalent to 2.3% of the Gross Domestic Product (GDP).[3] Fortunately, this is less than it was a few decades ago. For example, during the 1980s the hydrocarbons sector represented, in the peak of that decade, almost half (44%) of the State's annual income[4] and 12% of GDP.[5]

The reliance on hydrocarbons, then, originates from an economic perspective along with issues of national identity, and both cause difficulties regarding Mexico's energy transitions. Additionally, the lack of public policy and an international perspective also directly affects the energy transition in this important but polluting industry.

6.2 The Expropriation of the Oil and Gas Industry

The energy transition and the oil and gas industry in Mexico have a peculiar relation. This industry is strongly linked with politics, economy and national identity.[6] This issue is an obstacle for the energy transition. For that reason, it is worth explaining, even briefly, the historical and social context of the oil and gas industry in Mexico in order to place the

---

[2] Secretaría de hacienda y crédito público, 'Criterios Generales de Política Económica para la Iniciativa de la Ley de Ingresos y el Proyecto de Presupuesto de Egresos de la Federación correspondientes al ejercicio fiscal 2021' (Secretaría de Hacienda y Crédito Público, 2020), https://www.finanzaspublicas.hacienda.gob.mx/work/models/Finanzas_Publicas/docs/paquete_economico/cgpe/cgpe_2021.pdf, accessed 5 March 2021.

[3] World bank, 'Oil rent (% of GDP) - Mexico' (*World Bank*, 2018), https://data.worldbank.org/indicator/NY.GDP.PETR.RT.ZS?end-2018&locations=MX&start=1970, accessed 1 April 2021.

[4] Centro de estudios de las finanzas públicas, 'Estadísiticas Históricas de las Finanzas Públicas en México 1980–2002' [2002] 028(1) Centro de Estudios de las Finanzas Públicas, https://www.cefp.gob.mx/intr/edocumentos/pdf/cefp/cefp0282002.pdf, accessed 02 April 2021.

[5] World bank, 'Oil rent (% of GDP) - Mexico' (*World Bank*, 2018), https://data.worldbank.org/indicator/NY.GDP.PETR.RT.ZS?end=2018&locations=MX&start=1970, accessed 1 April 2021.

[6] Omar F Salinas González, 'Petrocultura y transición energética en México' (Nexos Blog del Medio Ambiente, 21st june 2021), https://medioambiente.nexos.com.mx/petrocultura-y-transicion-energetica-en-mexico, accessed 1 July 2021.

context of the complex relationship between this industry and the energy transition.

In 1938, the then President General Lazaro Cárdenas expropriated part of the hydrocarbons sector; previously, the hydrocarbons industry was totally liberalized, and private companies could participate without any restriction. In fact, foreign companies dominated approximately 95% of the total oil production.

The expropriation was a direct consequence of the observance or compliance of a Supreme Court of Justice's resolution that aimed to protect labour rights, due to a violation of employee's constitutional rights. When the resolution was notified, some of the international companies refused to abide by the final resolution.[7] As a consequence, President Lázaro Cárdenas decided to expropriate those companies. The decision was supported by most of Mexican society and enhanced the nationalistic sentiment. The political speech was that of the protection of both natural resources and sovereignty.

At the same time, President Lázaro Cárdenas with the Congress of the Union created PEMEX, published in the Official Journal for the Federation dated 20 July 1938. The purpose of this company is to operate the existing infrastructure, previously expropriated, and develop upstream, midstream and downstream activities.[8] But it was not until 9 November 1940 that article 27 was reformed, reserving the rights to exploit hydrocarbons to the nation and prohibiting the granting of any concessions.[9]

Nevertheless, for subsequent administrations, the hydrocarbons sector has been crucial for the development of Mexico. According to the World Bank. Nowadays—with a more diversified economy—this issue

---

[7] Pablo Escalante gonzalbo and others, Nueva Historia Mínima de México (12 edn, EL Colegio de México 2004) 469.

[8] Congreso de la Unión, 'Decreto de creación de Petróleos Mexicanos' [1938] 1(1) Diario Oficial de la Federación, http://dof.gob.mx/nota_to_imagen_fs.php?cod_diario=186811&pagina=3&seccion=0, accessed 3rd April 2021.

[9] Congreso de los Estados Unidos Mexicanos, 'Decreto que se adiciona el párrafo sexto del artículo 27 constitucional (petróleo)' [1940] 1(8) Diario Oficial de la Federación, http://www.dof.gob.mx/nota_to_imagen_fs.php?cod_diario=194840&pagina=2&seccion=0, accessed 25th March 2021.

has declined somewhat.[10] Despite the constant reduction in production, PEMEX remains one of the most important taxpayers for the Federal Government. In 2018, PEMEX alone represented 11% of the total income for the Federal Government,[11] equivalent to 2.3% of GDP,[12] and currently, PEMEX is one of the most important employers in Mexico. As of 2020, PEMEX reported 125,735 employees.

6.3 The Mexican Hydrocarbon Sector and the Energy Transition

6.3.1 Brief Overview Among the Most Relevant Reforms

For the purposes of this chapter, analysis of the hydrocarbons sector in Mexico will be divided into three milestone time periods: 1938 to 2014; 2014 to 2018; and 2018 until today.

6.3.1.1 1938 to 2014

Following expropriation of the hydrocarbons sector by President Lázaro Cárdenas and the creation of PEMEX as a National Oil Company (NOC), the Federal Government ruled the exploration and exploitation of the hydrocarbons as part of the Federal Administration but without the speciality required. Surprisingly, it was not until 1994 that the Mexican Ministry of Energy was created as we know it today.

[10] Yolanda Morales, 'Para el 2021, los ingresos petroleros de México representarían sólo 27% del PIB' (El Economista, 8 September 2020), https://www.eleconomista.com.mx/economia/Para-el-2021-los-ingresos-petroleros-representarian-solo-2.7-del-PIB-20200908-0003.html, accessed 30 March 2021.

[11] World bank, 'Oil rent (% of GDP) - Mexico' (*World Bank*, 2018), https://data.worldbank.org/indicator/NY.GDP.PETR.RT.ZS?end=2018&locations=MX&start=1970, accessed 1 April 2021.

[12] World bank, 'Oil rent (% of GDP) - Mexico' (*World Bank*, 2018), https://data.worldbank.org/indicator/NY.GDP.PETR.RT.ZS?end=2018&locations=MX&start=1970, accessed 1 April 2021.

Prior to 1994, the functions of the Ministry of Energy were incorporated into different ministries such as the Ministry of National Property and Administrative Inspection (1946–1958), Ministry of National Heritage (1958–1976), Ministry of Heritage and Industry Development (1976–1982) and Ministry of Energy, Mines and Public Industry (1982–1994).[13]

At the end of the 1970s and into the early 1980s, the Mexican economy depended almost entirely on oil and gas commercialization. The national economic crisis obliged the Mexican Government to depend on oil exportations. In 1980, for example, 72% of total sales came from the oil and gas industry, 76% in 1981 and 82% in 1982. Consequently, hydrocarbon taxation represented 25%, 26% and 44%, of the ordinary income for the Federal Government, respectively.[14]

The discovery of Cantarell's reservoir allowed PEMEX to become one of the largest producers, and the Mexican economy continues to depend on the exportation of hydrocarbons. Moreover, investment for the development and the incomes related to the commercialization allowed for the development of the regions in Mexico.

However, due to the oil and gas crisis in the 1980s and the country's important reliance on it, Mexico suffered one of the most important economic crises. Recovery of the oil prices did not happen until the 200s, when government investment in PEMEX permitted the company to reach its maximum production (3.4. million barrels per day).[15] Unfortunately, the received incomes were not used to relieve PEMEX debts; instead, it was used to boost the national economy.

The first reform to the oil and gas industry happened under the Administration of President Carlos Salinas de Gortari (1998–1994) with

13 Secretaría de energía, '¿Qué hacemos?' (Secretaría de Energía, 1st December 2018), https://www.gob.mx/sener/que-hacemos#:~:text=Congreso%20de%20la%20Uni%C3%B3n%20el,quien%20consider%C3%B3%20necesario%20lograr%20un, accessed 28 March 2021.

14 Francisco Colmenares, 'Petróleo y Crecimiento Económico en México 1938–2006' [2008] 5(15) Economía UNAM, http://www.scielo.org.mx/scielo.php?script=sci_arttext&pid=S1665-952X2008000300004, accessed 18 April 202.

15 Manuel Aguilera Gómez and others, 'Considerations on the Reform of the Oil Industry in Mexico' [2014] 11(33) Economía UNAM, http://www.scielo.org.mx/scielo.php?script=sci_arttext&pid=S1665-952X2014000300006, accessed 16 April 2021.

the aim to permit private investment in the gas market but only for commercial final users,[16] along with reform to the—nowadays abrogated—Law of Energy Electricity Public Services which allowed private parties to generate their own electricity.[17]

In 2008, President Felipe Calderón reformed the energy sector. The Federal Congress approved part of the initiative with the result of new laws aimed at addressing the climate change crisis and energy transition. Those laws were: (1) Law for Sustainable Energy Exploitation and (2) Law for Renewable Energy Exploitation and Energy Transition Financing.

The first law had the aim to oblige the Federal Government and its institutions to create a strategy, with clear aims and milestones to implement energy efficiency best practices, promote sustainable energy consumption, promote scientific and technological research, including educational programmes with a perspective on sustainable energy exploitation, and regulate energy efficiency and transport in the light of energy transition.

Further, the law also established the National Commission for Efficient Use of Energy (CONUEE)[18] as the technical body for sustainable energy use and in charge of promoting energy efficiency. This institution was one of the first governmental bodies specialized in energy efficiency in Latin America.

The second law looked at the promotion and use of renewable energy excluding hydropower with a generation capacity of more than 30 MW and nuclear power. It also obliged the government to promote sustainability, energy efficiency and the reduction of hydrocarbons as the primary energy source.

---

[16] Before the Energy Reform in 2014, the liquid petroleum gas importation and processing was exclusively for PEMEX.

[17] Congreso de la Unión, 'DECRETO que reforma, adiciona y deroga diversas disposiciones de la Ley del Servicio Público de Energía Eléctrica' [1992] 1(1) Diario Oficial de la Federación, http://dof.gob.mx/nota_detalle.php?codigo=4705440&fecha=23/12/1992, accessed 15 April 2021.

[18] The CONUEE was created by the same law on 28 November 2008.

Both laws were a starting point to legislate about energy transition and the development of a strategy for energy transition and energy efficiency. Nonetheless, the new legislation did not directly consider the hydrocarbons sector, specifically the activities developed by PEMEX for exploitation and exploitation. In other words, both laws failed to set clear pathways for the hydrocarbon activities developed by PEMEX. As with many countries around the world enacting similar legislation, the focus lay on different activities like electricity, argued to be the "low hanging fruit" of energy transition.[19]

6.3.2 Energy Reform 2014

During the administration of President Enrique Peña Nieto, the energy sector suffered the most important reform since the expropriation in 1938. The liberalization of the energy industry, a new legal framework, the creation of new governmental organizations and/or reform of existing ones revealed the then government's determination to renovate the sector. Nonetheless, for the purpose of this chapter, focus will be placed on the hydrocarbons sector.

The reform of article 28 of the Federal Constitution of the United Mexican States allowed the participation of private parties in exploration and exploitation activities as a petroleum operator or exploration company authorized through a contract or approval granted by National Hydrocarbon Commission (in Spanish the Comisión Nacional de Hidrocarburos (CNH)).[20] This reform also permitted private investment in both the midstream and downstream segments previously approved by the Secretariat of Energy (in Spanish Secretaría de Energia) (SENER),[21]

[19] Geoffrey Wood, Fossil Fuels in a Carbon-Constrained world, in Geoffrey Wood and Keith Baker (eds.), The Palgrave Handbook of Managing Fossil Fuels and Energy Transitions (Palgrave Macmillan, 2020), pp. 3–23.

[20] The National Hydrocarbons Commission's legal framework was reformed. The new regulation gave it independence and technical autonomy. Some of the faculties is to approve the Exploration and Development Plans and authorize annual budgets and new contractors, the National Centre for Hydrocarbon Information's management and sanction according to the law and/or contract.

[21] Gobierno de México, SENER Secretaría de Energia (Secretariat of Energy). https://www.gob.mx/sener/en.

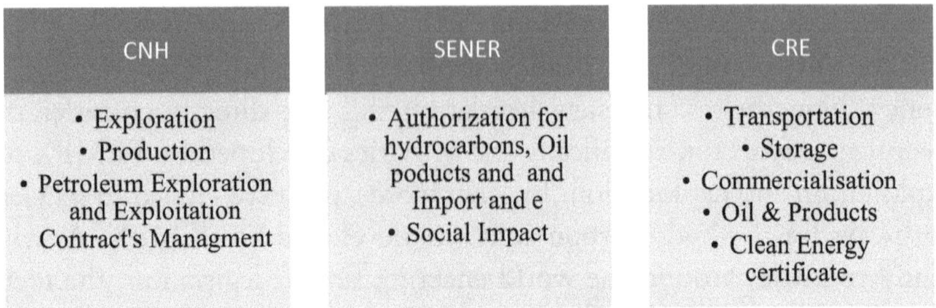

Fig. 6.1 Functions of CNH, SENER and CRE in Mexico

the government department in charge of the production and regulation of energy in Mexico and/or the Energy Regulatory Commission (in Spanish Comisión Reguladora de Energía) (CRE)[22] (see Fig. 6.1).

Once the Constitution was reformed, the Congress of the Union approved different laws to implement the constitutional mandate, including the Hydrocarbons Law, Climate Change Law, Law of the Electric Industry, Law of Coordinated Regulatory Bodies in Energy, Law of the National Agency of Industrial Safety and Environmental Protection for the Hydrocarbons Sector, among others. As the title notes, the Hydrocarbons Law regulated the entire hydrocarbons industry (upstream, midstream and downstream). It also entitled SENER, CNH and CRE to approve, audit, manage and sanction activities according to the legal framework.

At the same time, the Congress of the Union approved the creation of the Security, Energy and Environment Agency (in Spanish Agencia de Seguridad, Energía y Ambiente) (ASEA).[23] This Agency depends on the Ministry of Environmental and Natural Resources and is specialized in the hydrocarbons sector. The main role of ASEA is the approval of Environmental Impact Assessments (EIA), Industrial Safety programmes, the insurance register and to investigate environmental violations or accidents.

[22] Gobierno de México, CRE Comisión Reguladora de Energía (Energy Regulatory Commission). https://www.gob.mx/cre.

[23] Gobierno de México, ASEA Agencia de Seguridad, Energía y Ambiente (Security, Energy and Environment Agency). https://www.gob.mx/asea/

Due to the reform, the governmental organization gained importance when facing towards the participants in the hydrocarbons industry. Before the energy reform, besides environmental protection or permissions, the rest of the activities were practically unregulated. In fact, PEMEX did not require authorization for its E&P activities. On the other hand, CNH was a technical organism depending on SENER, in which its main role was to provide non-binding technical recommendations to PEMEX.

With the energy reform, the CNH becomes relevant. The technical and financial autonomy and the authorization laid down in the Law of Coordinated Regulatory Bodies in Energy allowed it to grant the E&P Contracts, authorize exploration and production plans, superficial exploration and recognition, oversee the CNIH, contract management, and PEMEX contractual or assignments areas technical management[24] (both from a technical perspective and compliance).

As a technical body, CNH has the obligation to assure the oil and gas industry's development, maximize oil and gas recovery, and ensure natural gas exploitation. Unfortunately, the energy transition is not part of its authorization. Nonetheless, there are regulations issued by this authority that indirectly contribute to the energy transition like natural gas exploitation with a minimum percentage to be complied with by PEMEX or each petroleum operator.

At the same time, CRE is in charge, besides the electricity sector, to regulate the storage, transport, distribution, liquefaction, regasification and commercialization of all hydrocarbons and oil products. Similar to CNH, however, there is no specific authorization regarding energy transition, although again indirectly, CRE can influence efforts towards energy transition via its technical regulations.

Moreover, the energy reform considered climate change and energy transition. In particular, the legal framework recognized the milestones

[24] Due to the reform, the round zero was exclusively for PEMEX. PEMEX decided which area is preferred to be maintained or developed. Once PEMEX chooses the areas, the rest of territory with prospective resources (oil & gas) were part of the following rounds 1, 2, 3 and 4.

proposed by the government at the COP21 that led to the Paris Agreement.[25]

For that reason, Mexico published the General Law of Climate Change that contributes to the legal framework previously established by the Law for Sustainable Energy Exploitation and the Law for Renewable Energy Exploitation and Energy Transition Financing. However, the regulation is too broad and public policy was focused on emissions reduction via renewable or clean energy.

Lastly, it is also important to note that the international legal framework holds relevance at the domestic level due to its obligatory nature according to articles 1 and 133 of the Political Constitution of the United Mexican States.[26] Thus, the Kyoto Protocol, Paris Agreement and the United Nations Framework Convention on Climate Change (UNFCCC) are part of the national legal framework and therefore binding. One of the main commitments is the reduction of greenhouse gas emissions of 30% in 2020 and 50% in 2050 against the baseline year of 2012. Unfortunately, the current administration does not really attend to such international commitments.

6.4 The Hydrocarbon Sector and the Energy Transition

As mentioned above, the hydrocarbon sector in Mexico has been crucial for national development even in the worst economic crises like the latest such episode in the 1970s and part of the early 1980s. Moreover, the implementation of social programmes, public policy and the federal annual budget depends on those incomes.

Additionally, it is important to consider that the electricity generation mix depends majorly on hydrocarbons. According to the IEA, in 2020

---

25 United Nations Framework Convention on Climate Change (UNFCCC), COP 21, https://unfccc.int/process-and-meetings/conferences/past-conferences/paris-climate-change-conference-november-2015/cop-21.

26 To be binding in Mexico an International Treaty, it is necessary to be ratified by the Senate and published in the Official Journal of the Federation.

natural gas and oil represent 63.4% and 8.4% of the energy mix, respectively,[27] of the resources for electricity production. In other words, the Mexican electricity generation matrix is completely dominated by fossil fuels.

Finally, another aspect to consider is that the oil and gas industry had been associated with national pride and national identity. Since the expropriation, the hydrocarbons have a strong correlation with political issues and are a central cause of economic growth and improvements in the quality of life standard among the population that is the basis of the legitimacy of a State monopoly and State property over the oil and gas reserves.[28]

With the current administration of President Andrés Manuel López Obrador, this ideology forms the basis of the energy policy combined then with a strong idea of sovereignty over the natural resources (oil and gas), energy sovereignty and energy security.

Those concepts have different definitions. This administration seems to have the obligation to develop the activities by the NOC (PEMEX) to reach self-sufficiency in order to reach energy sovereignty. To do this, it is necessary to strengthen the State companies even though it could have consequences.[29]

However, there is a misunderstanding between *sovereignty over natural resources*, *energy sovereignty* and *energy security* due to a lack of awareness of Resolution 1803 (XVII) of the United Nations General Assembly,[30] in which the concept concerning sovereignty over natural resources is defined as "*the right of peoples and nations to permanent sovereignty over their natural wealth and resources must be exercised in the interest of*

[27] IEA International Energy Agency, 'Electricity Generation Mix in Mexico' (IEA, 2020), https://www.iea.org/data-and-statistics/charts/electricity-generation-mix-in-mexico-1-jan-30-sep-2019-and-2020, accessed 17 April 2021.

[28] Alicia Puyana Mutis, 'Managing Mexican Oil: Politics or Economics? ' [2008] 16(32) Perfiles Latinoamericanos, http://www.scielo.org.mx/scielo.php?script=sci_arttext&pid=S0188-76532008000200004&lng=es&tlng=es, accessed 25 March 202.

[29] Energía a debate, 'Marca AMLO política energética por memorándum' (Energía a Debate, 4 August 2020), https://www.energiaadebate.com/regulacion/marca-amlo-politica-energetica-por-memorandum/, accessed 15 April 202.

[30] Mexico ratified the resolution; therefore, it is binding.

their national development and of the wellbeing of the people of the State concerned".[31]

The same resolution mentions that it involves "*the exploration, development and disposition of such resources, as well as the import of the foreign capital required for these purposes, should be in conformity with the rules and conditions which the peoples and nations freely consider to be necessary or desirable with regard to the authorization, restriction or prohibition of such activities*".[32]

On the other hand, energy sovereignty is a relatively new concept defined as the right of the citizens to choose the organization of energy usage.[33] In other words, "*energy sovereignty emphasizes the role of local people and their institutions in determining their energy systems in ways that are culturally relevant and ecologically sustainable*".[34] Instead, *energy security* is commonly defined as the reduction of vulnerability in the energy system. Its main features are availability, efficiency and environmental stewardship.[35]

The current administration mixed all those concepts with a nationalistic perspective or ideology. In its perspective, then, the government via its NOC (PEMEX) or the Federal Electricity Commission must develop all the activities to guarantee energy security and energy sovereignty even could affect efficiency, investment and projects according to international trends.

31 United nations, 'Resolutions adopted in the reports of the Second Committee 1803 (XVII) Permanent sovereignty over natural resources ' (United Nations, 14 December 1962), https://www.un.org/ga/search/view_doc.asp?symbol=A/RES/1803(XVII), accessed 1 April 2021.

32 United nations, 'Resolutions adopted in the reports of the Second Committee 1803 (XVII) Permanent sovereignty over natural resources ' (United Nations, 14 December 1962), https://www.un.org/ga/search/view_doc.asp?symbol=A/RES/1803(XVII), accessed 1 April 2021.

33 Chelsea Schelly and others, 'Energy policy for energy sovereignty: Can policy tools enhance energy sovereignty' [2020] 205(1) Solar Energy, file https://doi.org/10.1016/j.solener.2020.05.056, accessed 31 August 2021.

34 Murodbek Laldjebaev, Energy security, poverty, and sovereignty: Complex interlinkages and compelling implications. in Lakshman Guruswany (ed), International Energy and Poverty: The Emerging Contours (Routledge 2016) 97–112.

35 Murodbek Laldjebaev, Energy security, poverty, and sovereignty: Complex interlinkages and compelling implications. in Lakshman Guruswany (ed), International Energy and Poverty: The Emerging Contours (Routledge 2016) 97–112.

For that reason, the main decisions in the oil and gas sector are: 1) an increase in the oil and gas extraction by PEMEX and 2) to develop a new refinery (Dos Bocas). Both decisions are totally inconsistent with the energy transition and the milestones proposed by the Mexican Government during the COP21 Paris Agreement.

6.5 PEMEX Business Plan

PEMEX business plan for 2021–2025 does not contemplate an energy transition plan despite what other national or international oil companies have been doing during the last years—for example ENI, REPSOL, SHELL and ECOPETROL, among others.[36]

The business plan clearly focuses on gasoline import substitutions and reaches a misunderstood view of energy sovereignty. To achieve both aims, PEMEX has invested in a new refinery, presented a modernization programme of the five existing refineries and pretends to increase the oil and gas production. In fact, it has been increasing its production according to the information published by CNIH.

PEMEX has raised the production during the last few years but is far from the target established by the NOC. The business plan expects to produce 2,000 MBPD for this year, but according to the CNH, PEMEX reported a daily production of 1,606 MBPD.[37]

In the same document, PEMEX mentioned a reduction in its CO_2 emissions during 2017 and 2018 compared to 2015 and 2016. According to PEMEX, during those years PEMEX registered emission

[36] MaríaFernanda Ballesteros, 'Pemex, en las antípodas de la transición energética' (México Evalúa, 16 April 2021), https://www.mexicoevalua.org/pemex-en-las-antipodas-de-la-transicion-energetica/, accessed 2 May 2021.

[37] Centro nacional de información de hidrocarburos, 'Producción Nacional de Hidrocarburos' (Centro Nacional de Información de Hidrocarburos Estadísticas, 30 July 2021), https://hidrocarburos.gob.mx/media/4525/produccion-nacional-de-petroleo-y-gas-jul2021-1.pdf, accessed 27 August 2021.

levels below 40 million tons of CO_2 equivalent,[38] while in 2015 and 2016 they were above 50 million tons of CO_2 equivalent.

Despite mentioning about the reduction of CO_2 emissions in its business plan, the report published by the Think Tank *México Evalúa,* PEMEX Transformación Industrial, a subsidiary of PEMEX in charge of the refineries, petrochemical complexes and the processing centres, during 2018 and 2019, showed a rise in CO_2 emissions of 32% and SO_x emissions of 69%, respectively.

Furthermore, PEMEX trimestral gain report (July 2021) registered an incremental increase in CO_2 emission of 37% and SO_x emissions of 11% compared with 2020[39] due to natural gas flaring and infrastructural maintenance. The relevant issue is the incorporation of ESG metrics with the publication of the ESG information.

According to the ESG metrics, PEMEX is one of the riskiest oil and gas companies in the world from a corporate governance perspective in environmental, social and governance matters which affect the company's sustainability. The qualification in this metric is a minimum of 14.5 points and 59 points maximum. The higher the level, the riskier the company is. PEMEX punctuation is 56.7, only below Guanghui Energy (China) and Parsley Energy (United States of America) within 59.4 and 57.3, respectively.[40] However, there are oil and gas companies with good sustainability levels such as ENI and ECOPETROL with a punctuation of 25.7 and 37.3, respectively.

38 Petróleos mexicanos, 'Petróleos Mexicanos & productive subsidiary companies Business Plan 2018–2023' (PEMEX, 2018), https://www.pemex.com/acerca/plan-de-negocios/Documents/PEMEX_BUSINESS_PLAN_2019_2023.pdf, accessed 1st April 2021.

39 Bloomberg, 'Y hablando de Pemex aumentó 40% sus emisiones de contaminantes peligrosos' (El Financiero, 28 July 2021), https://www.elfinanciero.com.mx/nacional/2021/07/28/y-hablando-de-pemex-aumento-40-sus-emisiones-de-contaminantes-peligrosos/, accessed 28 July 2021.

40 México evalua, 'La responsabilidad corporativa de Pemex a prueba: El caso de los impactos sociales y ambientales de la refinería Miguel Hidalgo' (México Evalúa, May 2021), https://www.mexicoevalua.org/mexicoevalua/wp-content/uploads/2021/06/estudio-caso-pemex-tula-11-junio-final.pdf, accessed 5 June 2021.

6.6 Legal Framework for the Energy Transition

PEMEX does not have any clear pathway for the energy transition in its business plan. The Congress, CNH and ASEA have been publishing regulations for the oil and gas sector that contributes to the energy transition and will probably result in reducing greenhouse gas emissions if it's complied by PEMEX and the petroleum operators.

The oil and gas regulator published in the Official Journal for the Federation (in which all the regulations so published are binding) an Administrative Disposition for Natural Gas Utilisation. This regulation has the aim to determine the main features of the programme for natural gas exploitation. During the extraction programme, the target is the utilization of 98% of the natural gas. Nowadays, the amount of exploitation is around 88% according to the latest report presented by CNIH.[41]

Even this regulation encourages investment in infrastructure and reduces gas flaring, and the amount of natural gas used (not flaring) is far from the main target. From the author's perspective, CNH's lack of auditing and inspection visiting capacity as well as a deficiency in basic facilities and installations including pipelines, and the lack of a public policy to promote the investment in infrastructure are the main reasons that the operator does not comply with it.

The Mexican Government via the CENAGAS, CRE and/or SENER, in accordance with PEMEX and the other petroleum operators, is required to carry out public policy to enhance the development of infrastructure that allows natural gas exploitation and commercialization. Indeed, natural gas is seen as the fastest path to reduce CO_2 emissions in Mexico.

Another relevant regulation is the Administrative Dispositions that rules the prevention and integral control of methane emissions in the Hydrocarbons Sector published by ASEA in the Official Journal of the

[41] CNIH Centro nacional de informaci6n de hidrocarburos, 'Reporte de Aprovechamiento de Gas Natural Asociado' (Centro Nacional de Informaci6n de Hidrocarburos Estadísiticas, March 2021), https://hidrocarburos.gob.mx/media/4402/aprovechamiento-de-gas-natural-mar2021.pdf, accessed 15 April 2021.

Federation in 2018.[42] The purpose of this regulation is to set up the mechanisms and actions to prevent and control methane emissions only for (1) explorations and exploitation activities, (2) oil treatment and storage, and (3) natural gas liquefaction, transport, distribution, storage, regasification, compression and/or decompression.

According to EDF, "*methane has more than 80 times the warming power of carbon dioxide over the first 20 years after it reaches the atmosphere. Even though CO_2 has a longer-lasting effect, methane sets the pace for warming in the near term*".[43]

This regulation is mandatory for the activities mentioned above, but there are no sanctions in case of a violation. The absence of incentives, either punitive or nonpunitive, could affect its compliance. Nonetheless, it is a good initiative to reduce the oil and gas sectors' greenhouse gas emissions.

The regulation also allows third parties authorized by ASEA to inspect and certify the compliance of the Administrative Dispositions to make the procedure easier and reduce the regulatory and administrative burden.

6.7 Conclusions

From the analysis of the hydrocarbons sector with a special focus on PEMEX undertaken in this chapter, it can be concluded that there is not a consistent public policy or plan towards the energy transition. In fact, Mexico is trying to enhance the oil and gas industry in a direction away from the international trend.

PEMEX is one of the most pollutant oil and gas companies in the world. Nonetheless, it suffers from the lack of a clear pathway in its business plan about a transition to a lower carbon-intensive industry, and the energy transition is not mentioned or referred in any of its documents.

---

42 https://www.dof.gob.mx/nota_detalle.php?codigo=5543033&fecha=06/11/2018.

43 EDF Environmental defense fund, 'Methane: A crucial opportunity in the climate fight' (Environmental Defense Fund, 2021), https://www.edf.org/climate/methane-crucial-opportunity-climate-fight, accessed 20 August 2021.

A unified regulation and public policy is therefore necessary to enhance the energy transition. SENER should, as the authorized body in charge of energy policy, move to avoid the efforts of different governmental departments and organizations that delay or attempt to thwart addressing climate change and a low carbon energy transition. This is doubly important given that the energy transition could be a jobs generator even for the oil and gas industry. The International Energy Agency estimates that "*14 million new jobs are generated in energy supply by 2030. Over the same period, fossil fuel production could lose 5 million positions, resulting in a net gain of 9 million in this pathway*".[44] Mexico should adapt, accept and promote the new reality to recover part of the lost jobs during the pandemic.

[44] International energy agency, 'The importance of focusing on jobs and fairness in clean energy transitions' (IEA, 6 July 2021), https://www.iea.org/commentaries/the-importance-of-focusing-on-jobs-and-fairness-in-clean-energy-transitions, accessed 15 July 2021.

7

Energy Transition in Argentina: Past, Present and Future

Tomás Lanardonne and Juan Cruz Mazzochi

7.1 Argentina's Standing on the Energy Transition

7.1.1 The Global Agenda Impacts Argentine Energy Policy

Reducing the carbon footprint by 2050 has earned a major spot in the international agenda. So far, 196 countries have adopted the 2015 Paris Agreement, which provides for a coordinated effort to limit global

T. Lanardonne (✉)
Martinez de Hoz & Rueda, Buenos Aires, Argentina
e-mail: tomas.lanardonne@mhrlegal.com

J. C. Mazzochi
Martinez de Hoz & Rueda, Universidad Austral, Buenos Aires, Argentina
e-mail: jcmazzochi@mail.austral.edu.ar

G. Wood and J. F. Neira-Castro, *From Fossil Fuels to Low Carbon Energy Transition*, Energy, Climate and the Environment,
https://doi.org/10.1007/978-3-031-00299-1_7

warming. Signatory countries have agreed to maintain the global average temperature increase at 2 °C compared to pre-industrial levels[1] by means of a cooperative framework that sets out the main goals and targets regarding climate action. As provided under the Agreement, each party will create a 5-year plan under its own strategy and means of completion. Known as Nationally Determined Contributions (NDCs), this objective has driven the creation of new policies aimed at decarbonising the global energy sector, especially the extractive industries, power generation, heating and transportation, which are associated with traditional energy consumption. In short, energy transition is now fact and it has been further catalysed by the COVID-19 pandemic.[2]

Many public and private players have set in motion initiatives pursuing a zero-emission target. Technology has enabled unprecedented progress in the development and use of cleaner energy sources. However, questions remain on how, when and to what extent the energy transition will take place. Carbon sources have a long-standing tradition within the global energy matrix because they are a tried and tested energy source with a broad variety of uses (i.e. power generation, heating and transportation including aviation and ships). Moreover, even though many experts have forecasted a significant reduction over time in the use of fossil fuels, such energy sources still exhibit resilience in terms of cost, efficiency, power density and availability. In this respect, the biggest challenge that cleaner energy sources face is the ability to satisfy current and foreseeable energy uses in a cost-efficient way.

Nonetheless, global policymaking seems to set on reducing the carbon footprint. In order to achieve this, common ground must be reached in terms of a new energy paradigm. Traditionally, carbon, crude oil and natural gas have been associated with the notion of development. Over time, studies began to show that development can also be achieved by means of other energy sources with lower carbon emissions. In other words, new energy sources began to thrive as alternatives in segments where traditional energy sources were predominant (i.e. transportation,

---

[1] Pre-industrial levels refer to the period between 1850 and 1900.

[2] The recent US presidential election win by Joe Biden has resulted in the release of a clean-energy plan that includes a decarbonisation target by 2050 and carbon-free electricity by 2035.

industry, electricity generation). The energy portfolio became wider, with the focus being placed on finding efficient ways to meet the global energy demand in a sustainable way. Some forecasts display that by 2050, cleaner energy sources will coexist with 'sustainable' crude oil and natural gas. This scenario shows that decarbonisation is intended to happen by increasing the share of carbon neutral energy sources (i.e. wind, solar, green hydrogen) and by making efficient production and use of crude oil and natural gas. Accordingly, many oil companies have begun reshaping their strategy and directing efforts to the development of sustainable solutions. These include, carbon capture utilisation and storage (CCUS) systems, a shift towards natural gas, internal policies that prevent gas flaring and overall investments in technologies that improve the energy-efficiency indicator.

In Argentina, the commitment to reduce the nation's carbon footprint and go carbon neutral was ratified by means of the NDCs as part of the Paris Agreement.[3] This objective has also driven the enactment of several regulations, for example, Federal Law No. 27,520 which established minimum mandatory conditions for the adjustment and mitigation of global climate change, Federal Law No. 27,430 which created a carbon dioxide (CO_2) tax, and Resolution No. 339/2021 of the Secretary of Energy which implemented the use of Liquefied Petroleum Gas (LPG) as an alternative to fuel for sailing purposes.

7.1.2 Argentine Energy Mix: Past, Present and Future

A good starting point in order to understand Argentina's energy paradigm is to look at its energy matrix. Argentina has an energy mix[4]

[3] Argentina has presented two NDCs with short- to medium-term targets, that is, until 2030. The second NDC went even further than the first one and increased the original reduction commitment to 26% by 2030. The emissions shall not exceed 359 $MtCO^2$, while the last record shows that emissions are 364 $MtCO^2$.

[4] The total internal energy offer in 2018 was distributed as follows: 58% natural gas, 28% crude oil, 5% hydraulic, 3% renewables, 2% biofuels, 2% carbon and 2% nuclear. Data taken

made up mostly of natural gas, followed by crude oil. This matrix has a significantly small share of coal, and in the past years, renewable energies such as solar and wind have seen their share in the energy mix grow steadily. From a natural resources' perspective, Argentina has a broad range of available energy sources and reserves. Traditionally, Argentina has focused on hydrocarbons development, but recently, it has endorsed energy matrix diversification by means of an increase in renewable sources.

Another key indicator to consider is that of coal emissions. In spite of having a carbon-constrained matrix, Argentina generates an estimate of 8.6tn of carbon emissions per person which is average considering G20 emission rates.[5] According to official reports,[6] 53% of Greenhouse Gas (GHG) emissions are generated by the energy sector,[7] 37% by farming practices and other land uses, 5.5% by industrial processes and product use, and the remaining 4.5% by residues. Overall, this has exhibited a certain consistency since 1990. Therefore, even though there has been a significant increase in the share of renewables, levels of emissions per unit of energy have remained steady, showing only a modest decrease in the last few years.

Diving deeper into Argentina's energy mix, the main component is natural gas, making up over 50% of the total share. Many experts have referred to natural gas as the most resilient and cleanest traditional energy source, even placing it as the preferred hydrocarbon towards the 2050 benchmark: '*Near-term prospects for gas seem strong, for reasons reflecting*

from Secretaría de Gobierno de Energía, 'Escenarios Energéticos 2030. Documento de síntesis' (2019), http://www.energia.gob.ar/contenidos/archivos/Reorganizacion/planeamiento/2019-11-14_SsPE-SGE_Documento_Escenarios_Energeticos_2030_ed2019_pub.pdf, last accessed 10 April 2021.

[5] Climate Transparency, 'Brown to Green: The G-20 Transition to a Low-Carbon Economy' (Brown to Green Report, 2018), https://www.climate-transparency.org/wp-content/uploads/2019/01/BROWN-TO-GREEN_2018_Argentina_FINAL.pdf, last accessed 15 April 2021.

[6] Ministerio de Ambiente y Desarrollo Sostenible, 'Inventario Nacional de Gases de Efecto Invernadero y Monitoreo de Medidas de Mitigación' (2019), https://inventariogei.ambiente.gob.ar/files/inventario-nacional-gei-argentina.pdf, last accessed 14 April 2021.

[7] The emissions generated by the energy sector are distributed as follows: 33% energy industry (i.e. power generation, fuel production and oil refining); 26% transportation by land, sea and air; 18% other sectors including residential, commercial and agriculture use of energy; 17% construction and manufacturing; and 6% fugitive emissions caused by oil and gas flaring and venting.

the fuel's superior air quality attributes in comparison with coal or liquid fuels'.[8] This is why natural gas is called the 'bridging fuel' which is expected to lead the energy transition until renewables, hydrogen and other solutions scale-up.

Natural gas has been key in industrial, power generation and residential uses since the discovery of hydrocarbons back in the early 1900s. In 1992, with the enactment of the Argentine Gas Law No. 24,076, natural gas was given a place of strategic importance in Argentine policymaking. Most recently, in the last quarter of 2020, the Argentine Government launched a new stimulus plan called 'Gas Plan 2020–2024' in order to increase gas production to match local demand, given that imports are currently needed to satisfy domestic demand.[9]

In the light of the foregoing, Argentina could significantly contribute to the energy transition by being a global supplier of natural gas. Argentina has one of the biggest natural gas reserves in the world. The *Vaca Muerta* formation, located in Neuquén Province, is considered the world's second-largest shale gas deposit, and currently, less than 10% of its acreage is being developed through factory-drilling. The potential for low-cost development of natural gas could enable both covering local demand and exporting the surplus, which by its scale enables LNG developments.[10] Currently, '*Vaca Muerta already competes with the best U.S. tight oil basins in Texas and New Mexico from a well productivity perspective*'.[11] Numbers show that activity is recovering to pre-pandemic levels after several months of lockdown measures and overall depression of oil

---

[8] Akoz Losz and Jonathan Elkind, 'The Role of Natural Gas in the Energy Transition' (2019) Columbia | SIPA Center on Global Energy Policy, https://www.energypolicy.columbia.edu/sites/default/files/pictures/Role%20of%20Gas%20in%20the%20Energy%20Transition_CGEP_Commentary_092319-2.pdf, last accessed 5 April 2021.

[9] The 2021 natural gas demand is estimated at 46,850MM m^3. About 82% of demand is expected to be covered by local production and the remaining 18% by imports. Of the necessary imports to match demand, projections estimate that 10% will come from Bolivia and 8% from LNG imports. In winter, the import share is expected to increase to 26%.

[10] Faouzi Aloulou and Victoria Zaretskaya, 'Growth in Argentina's Vaca Muerta Shale and Tight Gas Production Leads to LNG Exports' (2019) U.S Energy Information Administration, https://www.eia.gov/todayinenergy/detail.php?id=40093, last accessed 2 April 2021.

[11] Rystad Energy, 'Argentina's Vaca Muerta Tight Oil Deposit Is Now Producing at Record Levels, Matching US Well Scores' (2021), https://www.rystadenergy.com/newsevents/news/press-releases/argentinas-vaca-muerta-tight-oil-deposit-is-now-producing-at-record-levels-matching-us-well-scores/, last accessed 1 April 2021.

prices. The boost on hydrocarbon demand and the steady rise in benchmark prices have resulted in a new record for onsite fracking activities. Vaca Muerta's potential is slowly becoming a reality and it could be a key player in the energy transition.

Finally, and with equal importance, Argentina holds vast reserves of so-called future of energy resources: lithium and hydrogen. These two resources have the potential for decarbonisation of the transportation industry, among others, and as explained in the following section, Argentina has conditions for achieving low-cost developments that enable a quick scaling-up in production. This is another reason why Argentina could be key in the global energy transition.

7.2 Scattered but Steady Efforts to Diversify the Energy Matrix

In the last decade, Argentina's energy policies have focused on the development of unconventional oil and gas resources, and on the promotion of renewables such as solar and wind, in order to diversify the energy matrix. Since the 2014 amendment of the Federal Hydrocarbon Law, which included regulation to foster the development of unconventionals, *Vaca Muerta* has garnered most of the attention from both Federal and Provincial governments.[12] Simultaneously, global advances in technology and the resulting improvements in the cost-efficiency indicator have created new opportunities for the development of wind and solar energy[13] and even unveiled the possibility to relaunch hydrogen projects. These drivers slowly began to shift the focus of energy policy. In fact, the Argentine government began to redesign the energy matrix by

---

12 As of 2014, Vaca Muerta's indicators have significantly improved by, *inter alia*, more than doubling the lateral length of the wells (from 3576ft to 7270ft), reducing well costs at around 40% and triplicating the fracture stages. In 2020, Vaca Muerta produced 38,289,635 crude oil barrels and over 60% of the country's natural gas production. Data taken from FDC Consultants, 'Analysis of Vaca Muerta Oil Wells 2014–2020' (March 2021 Update), http://fdc-group.com/eventos/actualizacion-del-reporte-analisis-de-los-pozos-petroleros-de-vaca-muerta-2014-2020/, accessed 1 April 2021.

13 For example, Argentina launched the 'RenovAR' and 'MATER' stimulus plans under which the wind and solar energy contribution to the power grid is estimated at 5012 MW.

creating new programmes seeking to scale-up renewable energy sources. The purpose of these policies was to diversify the matrix and direct consumption trends towards cleaner energy sources. In this section, we will provide insights into how Argentina is making the energy transition.

7.2.1 Renewables: Wind and Solar Power

Argentina's approach to renewables involves a mix of choices intended to apply simultaneously to different sectors and industries. Traditionally, biomass and hydroelectric energy developments have been a major part of the renewable share in Argentina. For example, the Yacyretá Dam, located in Corrientes Province, and the Salto Grande Dam, located in Entre Ríos Province,[14] together with other hydroelectric projects account for 17% of electricity generation in the country. But the catalogue does not end there. Argentina is also a top global producer of biofuels.

With regard to biofuels, Argentina is the fourth largest producer of biodiesel in the world. The combination of vast amounts of fertile lands, farming know-how, especially soybean cultivation, and the integration between feedstock producers and biodiesel producers have launched the country into the top rankings and enabled the export of large volumes to the United States and the European Union. This growth was enabled by a thorough regulatory framework that intended to promote biofuels production.

In 2006, Congress passed Federal Law No. 26,093, also known as the Biofuels Act, which established a mandatory 5% blend of ethanol in gasoline and biodiesel in diesel by 2010. Thereafter, several regulations including congressional statutes, executive decrees and Secretariat of Energy and Undersecretariat of Hydrocarbons regulations were enacted to promote the use of biofuels. Such regulations consisted mainly of (a) establishing mandatory blends with traditional fuels, (b) establishing official prices and allocations of ethanol, from both from sugarcane and corn grain, and biodiesel and (c) tax incentives. Most recently, Congress passed Federal Law No. 27,640 by means of which it created

[14] The Yacyretá Dam, which Argentina shares with Paraguay, and the Salto Grande Dam hold an aggregate estimate power output of over 5000 MW.

a new regulatory framework for biofuels but under the same principles and guidelines as the previous legislations.[15] Nonetheless, actors in the biofuel industry await a recovery of global markets, especially the European Union and the United States in order to resume international trading.

Notwithstanding the aforementioned renewables, in recent years interest has divested into solar photovoltaic and wind energy due to new technologies that lower overall costs. As explained below, these technological breakthroughs enhanced Argentina's long-term potential in solar and wind resources.

Aligned with the idea of diversifying the energy matrix and expanding installed capacity of electricity, former President Macri's administration launched several measures to stimulate renewable energy development. In this respect, two important factors must be considered. Firstly, Argentina has abundant solar and wind resources: constant sun in northwestern areas and constant wind in Patagonia.[16] This presents a unique opportunity to develop both wind and solar farms that can provide abundant clean energy to the power grid. Secondly, Federal Law No. 27,191, which amended Federal Law No. 26,190,[17] established progressive renewable energy shares in overall power consumption to be achieved by 2017, 2019, 2021, 2023 and 2025. As provided therein, renewables shall gradually accrue up to a 20% share of the total energy consumption by 2025.

In effect, the former administration launched two programmes that were extremely successful in capturing the 'renewable stake' in Argentina: RenovAR and MATER:

a. The RenovAR programme consisted of four tenders (RenovAR 1, 1.5, 2 and 3) for the purchase of electricity from renewable energy sources

---

[15] The scope and terms of Federal Law No. 27,640 (including the mandatory blends, underlying rationale and term) will be covered in future and specific publications.

[16] The winds in the Patagonia are in average 10 m/s and the solar irradiance in the northwestern areas reaches up to 7.4 kMWh/m^2 (daily sum).

[17] Federal Law No. 26,190 created a promotional regime for renewables back in 2007. Later, it was amended by Federal Law No. 27,191 which implemented new mechanics (i.e. tax benefits and the Renewable Energy Development Trust) in order to boost the renewable energy share.

by CAMMESA[18] on behalf of distribution companies and large users choosing such a purchase mechanism. The renewable projects thereunder included wind, solar photovoltaic, biomass, biogas and small hydroelectric developments. These tenders were created by a set of resolutions from the former MINEM and the former Federal Secretary of Energy. Under the procurement documents, both CAMMESA and MINEM were in charge of opening the submitted offers, analysing the technical and economic aspects and finally, awarding the most convenient projects. The award triggered a period in which the successful bidder was required to execute a renewable PPA, and also an agreement where the bidder was incorporated as a beneficiary of the FODER trust fund.[19] A total of 147 projects were awarded for 4,466.5 MW of power. Currently, RenovAR projects for around 2,300 MW of power have reached COD.

b. The MATER programme was created by means of Resolution MINEM 217/17, which regulated the mechanisms provided by Section 9 of Federal Law No. 27,191.[20] It established the applicable regulatory framework for PPAs between eligible generators and large electricity users. Also, it allowed eligible generators to apply for the assignment of dispatch priority applicable in cases of curtailment in the transmission grid associated with limitations in available transmission capacity. As of the end of 2020, about 563 MW of the total 1093.6 MW of power was dispatched under the executed PPAs.

RenovAR and MATER were successful programmes under which a considerable renewable power output was awarded and contractualised.

---

[18] CAMMESA stands for '*Compañía Administradora del Mercado Mayorista Eléctrico S.A.*'. It is the administrator of the Argentine power grid and the agent in charge of dispatch in the Wholesale Electricity Market (also known as MEM).

[19] The FODER trust fund is the *Fund for the Development of Renewable Energies* designed to act as a guarantee to CAMMESA's payment obligations under the PPAs, payment of the project's sale price or purchase price (under certain conditions the beneficiary may have the right to sell the project to the State or, as the case may be, the State may have the right to require the sale of the project) and provide overall financial support to renewable projects.

[20] Section 9 of Federal Law No. 27,191 individualises large users in the Wholesale Electric Market and large demands, both with an average consumption higher than 300 kW. It imposes the obligation to reach the projected targets by adjusting their individual electricity consumptions.

Moreover, in bidding round RenovAR 1, the submitted offers amounted to 6366 MW out of the 1000 MW originally tendered, and in RenovAR 2, the offers amounted to 9391.3 MW out of the 1200 MW originally tendered. This shows that renewables had a huge momentum and that Argentina presented good long-term development conditions.

Nonetheless, the Argentine economic crisis, the change of administration following the 2019 elections, the issuance of several capital markets and foreign exchange regulations that limited the capacity to secure financing and repay loans, and the COVID-19 pandemic outbreak hindered the course of renewables in Argentina. Projects that did not reach COD relied greatly on the timely execution of EPC contracts which were jeopardised by the unsteady macroeconomic situation and by the COVID-19 outbreak. The economic volatility and the high effort demanding negotiations with the International Monetary Fund diverted attention away from renewables. Thus, a comprehensive action addressing the status of RenovAR and MATER is still pending and the mid to long-term scenario has become uncertain. That being said, the uplift of financial obstacles and an overall improvement in economic circumstances may further relaunch these programmes and even create conditions for the tender of new projects. As clearly seen in the RenovAR bidding rounds, there is a huge interest in Argentina's renewable resources and technical factors have already been de-risked. Natural resources that are available year-round, trained human resources with the acquired know-how and the estimated low-cost development of green hydrogen are competitive advantages to look into as the macroeconomic context improves.

7.2.2 Hydrogen

Hydrogen currently displays huge momentum on the world energy stage. Experts consider it the cleanest, most secure and versatile energy source to lead the energy transition. When hydrogen is consumed or produced under certain methods, it does not generate carbon emissions. Also, it can be used in industry, buildings, power generation and transportation. In short, hydrogen offers the possibility to decarbonise certain sectors by replacing fossil fuels, or even lowering the carbon footprint by blending it with natural gas.

Moreover, hydrogen introduces the possibility to store power coming from renewable sources. Inconsistency in renewable energy generation translates into a mismatch in availability and demand. Hydrogen dispels the paradigm that renewable energy is unfit to provide a stable supply of energy to the power grid. In this context, renewable power can be transformed into hydrogen by means of a power-to-gas process. Thus, hydrogen provides the opportunity to store this energy in large amounts with high efficiency rates (i.e. with low associated energy losses). This breakthrough allows the deferral in use of renewable power, and consequently, it can be held in standby until its use becomes necessary to cover shortages or stabilise the power grid. Furthermore, compressed hydrogen can be easily shipped like LNG and this creates an unmatched opportunity for Argentina: the global export of wind and solar energy.

Several countries, including Germany, Japan, and China, have already created specific frameworks in order to stimulate hydrogen development and boost its usage in power generation, transportation and buildings. However, the hydrogen economy requires a major scaling-up and a robust strategy in order to reduce costs and enable green hydrogen development.

Argentina was one of the first countries in the region to participate in the ISO/TC 197 Hydrogen Committee since its creation in the late 1990s. In 2006, the Argentine Congress enacted Federal Law No. 26,123 which created a promotional regime for the development of hydrogen. As provided therein, hydrogen is defined as both a fuel and an energy vector. Thus, the Law seeks to capture all potential uses of hydrogen. Nonetheless, the Law requires an enforcement authority to implement certain aspects of the programme under the Law, including (i) draft the National Hydrogen Plan, (ii) manage the Hydrogen Stimulus Fund (which is meant to sponsor the National Hydrogen Plan), (iii) authorise any hydrogen-related activity for its use as fuel or energy vector and (iv) assist the Federal Executive in issuing further regulations. The Federal Executive has not yet issued implementing regulations to Federal Law No. 26,123, including, most importantly, the appointment of an enforcement authority. Hence, in practice many of the above-mentioned aspects of the plan are still pending.

Considering Argentina's potential in hydrogen and recent technological advances, a bill to amend Federal Law No. 26,123 is before Congress. This bill takes a modern approach considering that since the original Law was enacted, significant progress has been achieved in terms of cost-efficiency. The bill includes an overall improvement of the objectives and an explicit reference to recent discoveries such as the possibility to create a blend with natural gas, hydrogen storage, uses in the electric power industry, mainly generation, and uses in the transport industry, among others. If enacted, this bill would be the first step towards creating a modern hydrogen regulatory framework in Argentina. Still, even if the bill is passed, hydrogen would not have an autonomous regime with specific features in respect of production, transport, distribution and use. Natural gas regulations do not apply to hydrogen, and therefore, there is actually a legal vacuum in this regard.

Nonetheless, according to international studies,[21] Argentina is one of the countries best placed for producing hydrogen in terms of cost, available resources (i.e. wind and solar energy sources) and trained personnel. This presents an enormous opportunity to become a world-class producer, with the ability to satisfy both local demand and export to countries which are already implementing hydrogen use such as South Korea, Japan and Germany.

Additionally, Argentina has an extensive gas network and suitable infrastructure which could be used to transport, distribute and export hydrogen. This is important for two reasons: firstly, at low concentrations from 5 to 15%, hydrogen can be used in actual applications such as cooking or heating with no significant changes in the end-user equipment, and secondly, hydrogen can be extracted from natural gas and sent downstream for transportation and other uses.[22] The United Kingdom has begun running tests that include up to a 20% hydrogen share in

[21] IEA, 'The Future of Hydrogen. Seizing Today's Opportunities' (2019), https://www.iea.org/reports/the-future-of-hydrogen, last accessed 4 April 2021.

[22] M. W. Melaina, O. Antonia, and M. Penev, 'Blending Hydrogen into Natural Gas Pipeline Networks: A Review of Key Issues' (2013) National Renewable Energy Laboratory, https://www.nrel.gov/docs/fy13osti/51995.pdf, last accessed 5 April 2021.

natural gas,[23] and many countries like Australia and the United States are shaping similar plans. Considering the growing worldwide experience in testing the blend and availability of the required infrastructure, Argentina could be a leading country in the region by implementing relatively low concentration hydrogen quotas and even exporting through existing pipelines to Brazil, Chile and Uruguay.

That being said, long-term economic incentives are key to boosting hydrogen developments and gradually implementing existing hydrogen uses by means of the natural gas and hydrogen blend. In parallel, financial obstacles must be overcome in order to attract foreign capital and enhance R&D for ongoing hydrogen projects. Finally, international treaties and other soft law instruments may greatly contribute towards better regional integration and efficient policymaking in Latin America to develop the hydrogen economy.

7.2.3 Lithium

Lithium batteries have become a major part of our everyday life. They are mainly used to power electronic devices and vehicles, and to store energy. From cellphones and cars to high-scale storage facilities, lithium is essential in today's society. This has driven countries and global companies such as Tesla, Panasonic and Ganfeng Lithium, into finding cheap sources of lithium that enable efficient routing from extraction to the marketing of the manufactured good.

EVs are set to provide a serious alternative to traditional vehicles fuelled by fossil fuels. The transportation segment is arguably the most advanced and where EVs have increasingly become a reality. Furthermore, technology has enabled important breakthroughs that allow, for example, the use of electric buses in big cities.[24] Electrifying the transportation segment is a key component in reducing the carbon footprint.

---

[23] See HyDeplot, 'Hydrogen Is Vital to Tackling Climate Change' (2021), https://hydeploy.co.uk/, last accessed 10 April 2021. Australia and the United States are also considering beginning tests.

[24] IEA, 'Global EV Outlook 2020. Entering the Decade of Electric Drive?', https://www.iea.org/reports/global-ev-outlook-2020 last accessed 4 April 2021.

However, there are still many questions about EVs: will the generated electric power meet the demand for EVs? Is the distribution network enough to reach charging stations? Are there enough charging stations to cover EV recharging demand? Are EVs cost-efficient? Is there enough lithium to cover the Li-ion battery demand?

These basic questions show that even though EVs are a solid alternative, major efforts are needed to gradually electrify the transportation system. For instance, China created a comprehensive plan that endorsed EVs which included subsidies, tax exemptions, promotion of charging infrastructure and a mandatory EVs quota for vehicle manufacturers and importers.[25] The numbers are clear: China's share in the world on the road electric cars in 2019 was 47%, and electric buses at almost 50%.[26] This shows that economic stimulation may play a significant role in the transportation energy transition.

In 2021, the Argentine Government launched several initiatives to boost electromobility and EVs. These initiatives seek the allocation of foreign investment in the extraction of lithium and manufacturing of batteries and cars in Argentina. In short, these policies pursue the creation of a one-stop business: extraction and production of batteries in Argentina's territory.

From a natural resources' perspective, Argentina is part of the so-called 'lithium triangle', a cross-border shared spot with Chile and Bolivia where more than 67% of the world's proven lithium reserves are found.[27] Most of these resources are found in brines, i.e. large saline lakes which contain significant amounts of lithium in the water. In order to obtain lithium, the extracted saline water needs to undergo evaporation in large

[25] Columbia University in the City of New York | SIPA Center on Global Energy Policy, 'Guide to Chinese Climate Policy. Electric Vehicles', https://chineseclimatepolicy.energypolicy.columbia.edu/en/electric-vehicles, last accessed 4 April 2021.

[26] IEA (n 21).

[27] See Andrés López, Martín Obaya, Paulo Pascuini and Adrián Ramos, 'Litio en Argentina. Oportunidades y desafíos para el desarrollo de la cadena de valor' (2019) Secretaría de Ciencia, Tecnología e Innovación Productiva, https://www.argentina.gob.ar/sites/default/files/bid-litio-final.pdf, last accessed 3 April 2021.

pools until only the mineral salt remains. This residue is then treated in order to obtain lithium.[28]

Upstream numbers show that Argentina has the world's second-largest lithium resource and is 3rd among the world's lithium-producing countries.[29] As of 2020, there were 2 operating projects with ongoing expansions, 2 projects undergoing construction, more than 10 in the feasibility and preliminary economic assessment stage, and 6 in the advanced exploration stage.[30] A foreseeable rise in the demand and price of lithium may act as powerful drivers to ramp up the Argentine worldwide share of production.

In line with the foregoing, the MPD and representatives of the Catamarca, Jujuy and Salta Provinces have joined forces to create the Lithium Board and intended to offer a specific communication channel between the Federal Government and lithium-rich provinces. As provided under the Argentine Constitution, the provinces own the natural resources located in their territories, but Congress retains the power to enact a mining code that applies to the entire country.[31] But most importantly, the MPD has announced the drafting of a Federal Electromobility Law which is set to be submitted to Congress. Although the bill has not been made public, the MPD has stated that the Law will seek vertical integration by, *inter alia*, providing that batteries and EVs be manufactured in the country with as much local content as possible.[32] This Law is

[28] Ibid. The solar evaporation method is cheaper than mineral extraction from pegamatites. The estimated OPEX for mineral extraction are USD 4000 per ton while in evaporation, they range between USD 2500 and 3000 per ton.

[29] U.S. Geolocial Survey, 'Mineral Commodity Summaries. Lithium' (January 2021), https://pubs.usgs.gov/periodicals/mcs2021/mcs2021-lithium.pdf, last accessed 4 April 2021.

[30] Also, there are more than 40 projects in the early development phase. The operations in the Fénix and Olaroz mines have announced expansions to significantly improve capacity. Data in Ministerio de Desarrollo Productivo – Secretaría de Minería, 'Informes Especiales' (2020), https://www.argentina.gob.ar/informes-de-la-secretaria-de-mineria/informes-especiales, last accessed 3 April 2021.

[31] Section 75, subsection 12 and Section 124 of the Argentine National Constitution. However, Mining Procedural Codes are sanctioned and enforced by the provinces.

[32] Argentina's automotive industry has a solid track record, including significant exports to South America countries, and reached a production of 314,787 units in 2019. A wide range of local and international automobile manufacturers produce and sell vehicles in Argentina. Although production halted with the lock down measures adopted due to the COVID-19

expected to contain a Promotional Regime with tax and export benefits, an EVs public transport quota, a 'Green Bond' consisting of tax credits for those who purchase EVs and a special trust to stimulate the developments.

This purpose is aligned with Federal Law No. 24,196 of Mining Investments that established a specific investment regime for mining activities carried out within the provinces that adhered to such Law.[33] Pursuant to Section 8 thereof, mining projects governed by Federal Law No. 24,196 are granted 30 years of tax stability as of the date of the feasibility filing. This means that the total tax burden (covering in general the creation of new taxes, repealing of tax exemptions and raises of existing taxes) shall not be increased.

These measures seek an overall improvement of investment conditions in the lithium industry targeting not only the extractive segment but also the manufacture of batteries and auto industry. Public and private stakeholders are in dialogue for further amendments to the applicable regulations and even the creation of a specific framework for lithium.

Nevertheless, Argentina's vast lithium resource must be coupled with a proper regulatory framework that grants stability and is also flexible enough to adapt to a flourishing and changing market. Firstly, there is a need for a consensus among the provinces on how to approach the increasing demand and growing interest in Argentina's lithium. Reaching a unanimous regulatory framework contributes to the vertical integration sought by policymakers, mainly because collaboration between public and private actors in the operations is a must. Secondly, the lithium industrialisation plan should be designed in a coordinated, interdisciplinary and cost-efficient way. The experience of other countries has shown that these plans require interaction and joint efforts between the public and private sectors. Last but not least, a comprehensive approach is key for the development of lithium. Argentina has ample battery

outbreak, the indicators are now showing signs of recovery and producers are further analysing the possibility of increasing exports.

33 The Catamarca Province adhered by means of Provincial Law No. 4759, the Jujuy Province by means of Provincial Law No. 4695 and the Salta Province by means of Provincial Law No. 6712.

R&D, a settled and resilient auto industry and a growing interest in alternative means of transport. The assessment of these factors can provide valuable insights into how value can be properly added to lithium feedstock and be converted into batteries and other manufactures.

7.3 The Need for Comprehensive Policymaking

Today's challenge is to achieve energy security while reducing the carbon footprint. Projections show that in Argentina, the energy demand is expected to grow at a 2.5% annual cumulative rate by 2030. Current global policymaking trends suggest that this growth will be covered by an increase in the renewables share. In a carbon-constrained matrix, the quest for increasing the renewables share and stimulating hydrogen and lithium use requires a consistent and creative effort to design a tailored transition plan.

Firstly, sustainable oil and gas may be the starting point: as described above, the energy sector accounts for 53% of greenhouse emissions. Recent international trends in the oil and gas industry show that companies are embracing the energy transition and putting in place measures that reduce carbon emissions. Thus, adequate regulations based on the economic analysis of law may facilitate these initiatives by, for example, providing tax exemptions, deferrals, accelerated amortisation and exemption of import duties related to CCS technology. Locally, oil companies have begun implementing global policies related to going carbon neutral by 2050. These initiatives include preventing gas flaring, investing in the development of renewables and exploring new technology to streamline the O&G extraction and production process.

Secondly, as described in the previous section, Argentina has a robust position in renewables and an even greater potential of achieving considerable shares of renewable power generation. The RenovAR and MATER programmes have been key for the development of wind and solar photovoltaic, as well as other renewables. However, following the COVID-19 outbreak, many questions have arisen regarding Argentina's economic volatility and the financial feasibility of these projects. In other words, although renewables have grown considerably in the last years, there are certain remaining barriers that need to be sorted out in order to fully develop Argentina's renewable potential.

Thirdly, Argentina could contribute significantly to the global energy transition by providing natural gas, lithium, either as feedstock or as batteries, and hydrogen. On the one hand, energy transition is said to be a process with a clear benchmark in 2050: going carbon neutral. Despite this ambitious goal, energy demands ought to be met by actual resources while providing a window for the scaling-up of alternative energy resources. Under these conditions, natural gas is the best fit and Argentina has vast resources of it. The *Vaca Muerta* formation, which has been already majorly de-risked, has the potential of providing a stable supply of natural gas to the world. In this respect, LNG developments may play a significant role in the export of natural gas to countries such as Japan, India and China.

On the other hand, technological breakthroughs relating to hydrogen and lithium have enhanced Argentina's potential over alternative energy resources. The conditions in Patagonia and the northwest of the country enable scaling-up at a competitive cost since abundant wind and solar power are available to produce green hydrogen. Likewise, the abundant gas resource enables alternative uses of extracted natural gas such as blue hydrogen production with CCS. Further, as seen in other countries, hydrogen can be gradually blended with natural gas into pipelines and end-user equipment with few modifications. In short, hydrogen may be a game changer for local users and it may also be exported to other countries which have already implemented its use in scale.

With regard to lithium, the foreseeable demand and rapid price increase present excellent conditions to deepen commitments in the northwest region where the reserves are located. Financial obstacles can be sorted through regulations that grant stability and legal remedies to investors. Also, the Argentine Government's electromobility initiatives may provide the opportunity to develop an internal electric vehicle market. In summary, although in different stages, these developments may flourish in Argentina since it has the necessary natural resources, trained human resources and investigators, and the oil and gas and renewables know-how that may be applied to these developments.

8

The Important Role of Mining Within the Energy Transition: The Case of the Lithium Sector in Argentina

Florencia Heredia, Agostina L. Martinez, and Valentina Surraco Urtubey

8.1 Introduction

Recently, there has been an increasing demanding for greenhouse emission-free products and services, ultimately leading the way to the rise of electromobility and cleaner technologies for sourcing energy. Under this scenario, the on-going energy transition is driving the demand for a new category of minerals named 'battery minerals' (lithium, nickel,

F. Heredia · A. L. Martinez (✉) · V. S. Urtubey
Department of Energy and Natural Resources, Allende & Brea Law Firm, Buenos Aires, Argentina
e-mail: amartinez@allende.com

F. Heredia
e-mail: fheredia@allende.com

V. S. Urtubey
e-mail: vsurraco@allende.com

G. Wood and J. F. Neira-Castro, *From Fossil Fuels to Low Carbon Energy Transition*, Energy, Climate and the Environment,
https://doi.org/10.1007/978-3-031-00299-1_8

cobalt, manganese, graphite, rare earth elements (REE), copper, and aluminum, among some others), for which the electric vehicle (EV) industry appears as a key driver.

The link between energy transition and the mining industry is very often overlooked by civil society, especially in Latin America. In this respect, mining is a natural condition precedent to the energy transition, meaning that without mineral resources no energy transition can be achieved. For example, a typical electric car requires six times the mineral inputs of a conventional car.[1]

Consequently, the mining industry will need to adjust their operations toward climate-smart mining policies and move toward using renewable energy sources, from a strict Environmental, Social, and Governance (ESG) perspective. Even though there are a variety of related aspects to be analyzed by the literature on the above topic, this chapter will address how mining can contribute to this energy transition, namely which one of the alternatives is based via the supply of a particular battery mineral: lithium. Specifically, we will focus on the case of Argentina and its lithium mining sector.

This chapter starts by introducing the basic concepts of lithium such as its uses, demand, and sources. Subsequently, the analysis will focus on the Lithium Triangle (encompassing Chile, Argentina, and Bolivia) with a particular focus on the case of Argentina. In order to better understand the legal and regulatory framework of Argentina in governing the lithium mining industry, the Argentinean constitutional organization will also be discussed.

Thereafter, a brief summary of the history of the lithium sector in Argentina is described, together with a description of the current lithium projects either being developed or at the exploration stage. Additionally, a description of certain initiatives placed or to be placed by the Argentine Government in relation to lithium are depicted. Finally, conclusions containing our main thoughts are provided.

---

[1] International Energy Agency, *"The Role of Critical Minerals in Clean Energy Transitions,"* p. 5 (2021).

8.2 Lithium in Context

Lithium is considered by the United States Geological Survey (USGS) as the "lightest of all metals" and, in its pure elemental form, it is a soft and silvery-white metal, highly reactive (for being an alkali metal) and therefore cannot be found as a pure metal in nature. Generally, lithium is combined with other minerals such as silicon, phosphorus, and oxygen.[2]

One of the most valuable and recognized increasing use of such a mineral is as a component of batteries and rechargeable lithium-ion batteries due to its use in electronic technologies and green industries, such as electric vehicles (EV), and smart phones, among others.[3]

The main chemical compounds which result from lithium extraction are: (i) Li carbonate; (ii) Li hydroxide and chloride; and (iii) metallic lithium. Li carbonate is usually the first step to the rest of the chemical compounds. It is important to note that lithium carbonate is not traded on international stock exchange markets, and therefore its price is dependent on direct negotiations between producers and end-users. In the past decade, its average value has increased fivefold, with a particularly significant increase in the past few years.

8.2.1 Lithium Demand

The demand for lithium is directly linked to the forecasted rise in EV sales. As demand grows, prices also fall. To illustrate such a notion in accordance with Bloomberg New Energy Finance's EV Outlook Report 2021, lithium-ion battery pack prices fell 89% from 2010 to 2020, with the volume-weighted average hitting US$137/kWh, and in this regard underlying material prices will play a larger role in future. Roskill's

---

[2] Bradley Dwight, and Jaskula Brian, *'Lithium—For Harnessing Renewable Energy: U.S. Geological Survey Fact Sheet'*, p. 2, (2014), see https://dx.doi.org/10.3133/fs20143035, last accessed on June 12, 2020.

[3] Andrés López, Martín Obaya, Paulo Pascuini, and Adrián Ramon, '*Litio en la Argentina: Oportunidades y desafíos para el desarrollo de la cadena de valor*', p. 11, Banco Interamericano de Desarrollo – BID – (2019); Gunther Martin, Lars Rentschb, Michael Höckb, and Martin Bertaua, "Lithium Market Research – Global Supply, Future Demand and Price Development" (2017) 6 Energy Storage Materials 171–178.

analysts estimate that lithium carbonate equivalent demand will rise above two million tons by 2030, representing a more than 4.5-fold increase from 2020.[4]

The forecast for EV deployment in future is quite significant. In accordance with Bloomberg New Energy Finance, it is largely anticipated that there will be an EV-dominated fleet by 2050, with almost 90% of cars sales being EVs. These numbers are particularly optimal for the lithium market. This notwithstanding, it is necessary to consider that technologies in the batteries sphere is an evolving matter, and it is not yet certain which battery chemistry will prevail in future. Still, battery mineral industry insiders usually claim that whatever the battery chemistry will be in future, lithium will also be a present component and one that will be indispensable.

8.2.2 Lithium Sources

Two of the primary sources of lithium are through the extraction of brines or "hard-rock." Brine extraction is today generally considered the most cost-efficient process. Most of the brine lithium resources are mainly located in Chile, Argentina, and Bolivia (the so-called Lithium Triangle), and also in China, Russia and the state of Nevada in the United States. Due to a variety of reasons, Argentina could be considered as the main investment target for many international investors, particularly considering the fact that—within the Lithium Triangle—the regulatory framework provides friendlier scenarios for investment. Moreover, in terms of geology, the Argentinean lithium region is to be considered as unexploited since only two lithium mines have been and are currently in production, as will be discussed below.

[4] Reuters 2021, "*Shortages Flagged for EV Materials Lithium and Cobalt,*" see https://www.reuters.com/business/energy/shortages-flagged-ev-materials-lithium-cobalt-2021-07-01/, last accessed on October 13, 2021.

8.3 The Lithium Industry in the Lithium Triangle

Chile, Argentina, and Bolivia (the Triangle States) in combination contain 49.9 million tons of lithium resources.[5] Notwithstanding the identification of the Triangle States as a whole, it should not go unnoticed that their policies and idiosyncrasy toward the mining sector—and in particular their lithium resources—completely differs one from the other.

Countries that have shown a long-term and steady mining policy have seen many positive results. Chile is certainly the best example of this, while Argentina and Bolivia have a very different situation, although they are on their way to becoming relevant lithium producers, despite some critical challenges to be faced in order to develop and operate a lithium project especially from a technical perspective, together with maintaining a long standing "social license to operate" (SLO).

On the one hand, in relation to the legal and regulatory framework, Chile considers lithium as a "strategic resource,"[6] non-susceptible to mining concessions to third parties; however, private sector participation in the case of the Atacama salar is allowed. In the case of this salar, private companies[7] actually operate and extract lithium by means of lease agreements granted by the owner of the mining lithium concessions within such salar. On the other hand, Bolivia also considers lithium as a strategic resource.[8] However, it has reserved extraction, industrialization, and commercialization rights to the state.

[5] United States Geological Survey, "Mineral Commodity Summaries," (2021) see https://pubs.usgs.gov/periodicals/mcs2021/mcs2021-lithium.pdf, last accessed on October 13, 2021.

[6] 40 Issued by the Chilean government on October 1979.

[7] 41 *Sociedad Química Minera* (SQM) and *Albarmale* (formerly Rockwood).

[8] 42 Sectoin 396 of the Political Constitution of the State of Bolivia, Section N° 396 (February, 2009).

8.4 Lithium in Argentina

Argentina is one of the few countries in a position to consolidate major reserves of lithium. According to the US Geological Survey, Argentina's lithium resources could be the largest worldwide, at 19.3 million tons and 2 million tons in reserves. Salars in Argentina are distributed in the provinces of Salta, Jujuy, and Catamarca (the Puna Region). Currently, there are only two salars in production: Salar de Olaroz (Jujuy); and Salar del Hombre Muerto (Salta and Catamarca).

Lithium extraction is regulated under national, provincial, and local regulations. Extraction of this mineral is very recent, with the first operational mine (in terms of extraction) commencing only 20 years ago. Despite this, it has attracted investment for quite a long time due to the low-cost salar brine extraction and its regulatory framework in comparison with other countries within the region. In order to obtain a better understanding of the mining regime which is applicable when extracting lithium, this chapter will provide an overview of the Argentine constitutional organization for lithium extraction and exploitation.

8.4.1 Argentinean Constitutional Organization

Argentina is a federal republic consisting of twenty-three autonomous provinces and the Autonomous City of Buenos Aires, organized under a national constitution (ANC). This fact plays a crucial role in any discussion of natural resources since these are located within the territory of the country's provinces. As mentioned above, in the case of lithium projects, these are mainly found in the Puna Region.

The national government of Argentina is organized into three separate branches. The executive branch is headed by a democratically elected President. The National Congress is comprised of a bicameral legislature, the House of Deputies and the Senate, with the Deputies and Senators also democratically elected. Finally, the judicial branch is led by the National Supreme Court of Justice, which is comprised of five justices appointed by the President subject to Senate approval. Beneath the Supreme Court is a hierarchy of federal courts. Argentina's legal system is civil law based, meaning that its courts rely on laws, mostly compiled in codes, rather than on precedent established in prior judicial decisions.

Due to the federal organization, each of the provinces has its own government and courts, though in certain areas provinces would also be bound to federal law (however, these boundaries are sometimes unclear). The ANC establishes that each of the provinces determines their own local institutions by which they will be governed. The national government may not intervene in the election of their governors, legislators, and other provincial officers. Finally, each province enacts its own provincial constitution and legislation.

According to the ANC, the provinces are the original owners of all natural resources existing in their territories. Therefore, excepting certain legal considerations, the main impact on the mining sector is that the provinces have the power to rule and decide on specific policies related to the mining industry, even within the scope of a federal or national resources policy.

8.4.2 Lithium Regulation: General Mining Legal Framework

Lithium as a mineral—and consequently its extraction and the development of lithium projects—is regulated as any other mining activity of other types of minerals and therefore is subject to the general Argentine mining and environmental legal framework. This means that Argentina lacks a specific regulation for the development of lithium projects. Moreover, the provisions of the Argentine Mining Code (AMC) and procedural provincial regulations apply to lithium and the granting of concessions. It is notable that the National Congress enacts the substantive mining legislation (as the AMC), while the provinces are empowered to regulate the procedures related to it (which cannot conflict with substantive legislation) and enact complementary regulations, which includes environmentally protected areas. Municipal governments have jurisdiction in regulation-related issues. Therefore, the mining industry is mainly regulated by the AMC, as amended, which governs the rights, obligations, and procedures regarding the exploration, exploitation, and use of mineral substances. The AMC sets forth the legal framework that rules the relationship between the state and those who develop any kind of mining activities (through an exploration permit or a mining concession).[9]

As mentioned previously, the provinces own the natural resources existing within their territories, although they are not allowed to exploit such resources directly. Therefore, provinces must grant mining rights to those individuals/legal entities interested in mining exploration and exploitation, the so-called right of "mining property" by means of a legal concession.[10] Once a mining concession is granted, the title holder owns all the mineral deposits within the boundaries of the property, irrespective of the mineral substance contained therein. From a different standpoint, the categories of mineral substances foreseen by the AMC can be summarized as follows: (i) those that belong to the State(s) and not to the surface landowner (i.e., mines corresponding to the first and second categories); and (ii) those that belong to the owner of the surface land in which they are located. It is further notable that lithium deposits are considered substances of the first category, subject to concession to be granted by the relevant province(s) to third parties.

8.4.3 Lithium History in Argentina

Mining industry at large scale only started in the 1990s when the Argentine government implemented an aggressive strategy to open the mining sector to foreign private investment, together with a liberalized

[9] In Argentina, the AMC provides for two types of mining rights—exploration permits and mining concessions—both of which are granted on a first-come, first-served basis. Exploration permits are exclusive authorizations to explore a certain area during a limited period and to the extent provided by the AMC, while mining concessions grant the title holder the right to conduct further exploration works after a discovery has taken place and to exploit all mineral deposits within the boundaries of the mine. Mining concessions are not subject to a life term and, therefore, to the extent the title holder does not incur in any of the concession termination events set forth in the AMC, the concession will last until the mineral reserves are exhausted.

[10] Please note that the original public domain pre-exists the mining property granted, coexists during the concession and subsists, still remains and continues even after the concession becomes void.

economy which took place in the country during the same decade.[11] The promotional mining regime was implemented by means of the Mining Investment Law Nr. 24.196 (1993), the Federal Mining Agreement (1993) and legislation which amended the 1887 AMC. Consequently, with a satisfactory promotional regime, initially mining investment took place mainly in the metalliferous mining sector.

Commercial lithium extraction operations in Argentina started in 1998 with the activities of the Fénix project owned by FMC Corporation—today Livent Corporation—at the Salar del Hombre Muerto, located in the province of Catamarca.[12] The Fénix project, together with the well-known metalliferous project "*Bajo La Alumbrera*," kick-started large-scale mining in the country, with both located in the province of Catamarca. The Fénix project was for many years the only lithium production mine in place in the country until 2015 with the first sale of lithium carbonate from the Olaroz project, operated by the joint venture between Orocobre Limited (now Allkem Limited, following a merger between the two Australian companies Orocobre Limited and Galaxy Resources Limited which took place in August 2021), Toyota Tsusho Corporation and a provincial state company called *Jujuy Energía y Minería Sociedad del Estado* (JEMSE).

In addition, there are a number of exploration projects with potential for development in the next few years. These vary in size and technical conditions, and it remains to be seen how many will eventually become feasible, and to what extent. There are a myriad of challenges (social, technical, and economic factors) playing a role in the competitive development of these projects, together with the general challenges that the mining sector is currently facing all over the world.

Quite apart from the availability of resources, there are many other factors playing a role in the competitive development of these projects. Such factors relate to: technical expertise; methods of extraction; the

---

[11] Bastida, Elizabeth, Irarrázabal, Ricardo Labó, and Ricardo (2005): "*Mining Investment and Policy Developments: Argentina, Chile and Peru,*" see https://www.researchgate.net/publication/277952408_Mining_Investment_and_Policy_Developments_Argentina_Chile_and_Peru, last accessed on October 13, 2021.

[12] Livent Corporation Form 10-K (2019), p. 11, see https://s22.q4cdn.com/453302215/files/doc_financials/2019/annual/10-K.pdf, last accessed on October 13, 2021.

infrastructure needed to bring the project to the distribution stage; and relations with the Puna Region local communities, which may differ from province to province or even within the same province.

The mining industry also faces big challenges all over the world, and lithium projects are certainly no exception. With projects located in remote areas of the Puna Region, the interaction among mining companies, governments (including provincial or state-owned mining companies), and ESG matters (especially in relation to social matters with communities) will be crucial. Table 8.1 summarizes the details of some of the current lithium projects under different stages in the country.[13]

In the past years, there has been an important rate of foreign direct investment in the lithium projects in Argentina, mainly from Australia, Canada, and more recently China, which has shown a clear strategy to advance toward high-end manufacturing, production of electric cars and clean energy technology as part of its policy. It is expected that China's role in the lithium sphere will continue to increase.

There are several projects with different operators which are located within a single salar. Additionally, there are salars that are located in the shared and adjacent territories of more than one province or jurisdiction. These two situations provide extra layers of complexity to the development of lithium projects since water and fluids know no boundaries and flow freely. Moreover, there is a lack of regulation of these complexities which will need to be addressed in the near future.

8.5 Specific Regulation for Lithium?

Given the very particular nature of lithium in salars, there are many technical hydrogeological aspects related to its extraction which need to be considered in a specific way. As described above, lithium mining activity is regulated by the provisions of the AMC which was enacted

---

[13] We note that the information is based on sources published by the Argentine National Secretariat of Mines, available at: https://www.argentina.gob.ar/sites/default/files/litio_en_argentina_-_wilson_center_ingles.pdf.

Table 8.1 Current Lithium projects in Argentine

Operation				
Province	**Project**	**Company**	**Capacity (2019)**	**Operation start year**
Catamarca and Salta	Fenix (Catamarca)	Livent	22,500 t LCE (expansion estimated at 28,500 t LCE)	1998
Jujuy	Olaroz (Sales de Jujuy)	Orocobre, Toyota and JEMSE	17,500 t LCE (expansion estimated at 42,500 t LCE)	2015
Construction				
Province	**Project**	**Company**	**Capacity (2019)**	**Operation start year**
Jujuy	Cauchari-Olaroz	Lithium Americas, Ganfeng and JEMSE	Estimated at 40.000 t LCE	2021
Salta	Centenario Ratones	Eramet	24,000 t LCE	2021 (currently suspended)
Exploration				
Province	**Project**	**Company**	**Capacity (2019)**	**Stage**
Catamarca	Sal de Vida	Galaxy Resources Ltd (Allkem Limited)	25,000 t LCE	Feasibility study undergoing Pilot plant
Salta	Sal de los Angeles	NextView New Energy Lion HK Ltd (Tibet)	25,000 t LCE	Feasibility study undergoing Pilot plant

(continued)

Table 8.1 (continued)

Operation				
Province	Project	Company	Capacity (2019)	Operation start year
Salta	Rincon Salar	Rio Tinto	25.000 t LCE 50.000 t LCE	Feasibility study completed Demonstration plant
Catamarca	Tres Quebradas	Neo Lithium (Zijin)	20,000 t LCE	Feasibility study undergoing Pilot plant 99.1% LCE
Salta	Sal de Oro	Posco Corporation	25,000 t LCE	Advanced exploration Pilot plant undergoing
Salta	Pozuelos—Pastos Grandes	Lítica Resources (Plus-petrol)	25.000 t LCE	Preliminary economic assessment in process

in 1886 and is a code designed to regulate primary metalliferous mining projects. Therefore, when applied to lithium mining activities, some gaps are inevitably encountered. In this sense, the Argentine Government has been anticipating discussions toward the analysis of potential specific regulation that could encompass all of the above mentioned aspects.

This process will require specific technical knowledge from the government at all levels. It also anticipates, firstly, the interaction between future developments and current plans for infrastructure development, and the economic and production agendas of the provinces involved. Also necessary for discussion is the role and scope of interaction to be played by public provincial mining companies and their participation in lithium projects—for example, in the case of Jujuy, lithium is considered a strategic mineral and this fact has several implications.

The best laws and regulations, when considering their application and enforcement, have proven to be those where consensus among all stakeholders is reached (or at least considered during the process). In this regard—and especially considering the many technical aspects that relate to the operation of lithium mines—in order to assure environmental protection and balance of the hydrogeological conditions, as well as other ESG concerns, a thorough assessment should be made. Similar regimes, such as those in the USA and Chile, could be worth taking into account in terms of guidelines and comparative experience. It is to be expected that governments, communities, and industry players work together effectively in order to develop certain general guidelines relating to the lithium sector. Within these guidelines as a general framework, specific protocols for each province and more specifically for each particular salar and its operation, could be developed. This could be a way to integrate general parameters with the situation of each project taking into consideration the relevant aspects and issues at stake.

8.6 The Argentine Lithium Region

Recently, on October 5, 2021, the governments of the provinces of Jujuy, Salta, and Catamarca subscribed an Interprovincial Treaty by means of which a mining lithium region was created (the 'Lithium Region'). Under this Treaty, a "Regional Committee of Lithium" was created and integrated with the main authority of each provincial mining authority, together with a representative of the National Ministries of Interior, Productive Development and Science and Technology (which depends on the Executive National Power). The functions of the committee are to act as an entity that shall coordinate the provincial requirements (either between the provinces and before the National Government) in relation to aspects of regional interest such as research, production, industrialization, and commercialization of lithium throughout the value chain. Finally, the Interprovincial Treaty will also be ratified by each provincial legislature and the National Congress. It is important to highlight that this interprovincial agreement between provinces, creating a "mining region" is the first of its kind in the country and arguably promises to be the first step in the consolidation of a stronger relationship among the provinces and thus fostering regional economic growth.

8.7 Argentinean Electromobility Initiative

As mentioned above, there is a widespread notion in Latin America that historically the region has been a mere provider of raw materials to developed countries and as such has undermined its own development. In this sense, there is a clear trend of mostly civil society demanding that the Lithium Triangle States should not be only the provider of the raw materials for the energy transition, but also take a more primary role and thus active role in the manufacturing of batteries. As of today, this has been materialized by means of different initiatives, encouraged by the various states. In this regard, stances are divided between those who believe that battery manufacturing is possible and even competitive in the Triangle States, and those who consider these initiatives as hard to achieve due to several factors including a lack of technical resources and/or know-how, along with a lack of access to the other elements needed for the manufacturing of batteries.

In the case of Argentina, and for the past years, there has been an intention of the Argentinean Government to enact a piece of legislation which regulates electromobility, among other matters. Recently, in 2021 the current administration announced the submission before the National Congress of a bill of law with the aim to create a Support Program for Electromobility, which set forth certain incentives and benefits to foster private investment to develop the industrialization and manufacturing of lithium and EV within the country.[14] In accordance with Section 2 of the Bill of Law, the purpose of such program is "to encourage the development of electromobility, in order to accelerate the transition to a low-carbon economy and ensure sustainable technological progress. The creation, design, development, production, construction, assembly, manufacturing of items, components and elements of electric vehicles will be promoted, as well as all the accessory processes, such as prototype design, patents and industrial designs."

[14] Please see https://www.telam.com.ar/notas/202108/565916-estiman-inversiones-por-us-900-millones-por-la-futura-ley-de-movilidad-sustentable.html, last accessed on October 13, 2021.

Considering that this Bill of Law is still subject to the legislative process, it remains to be seen whether or not the support scheme will foster electromobility; however, this should be considered as a first step.

8.8 Conclusion

The great lithium potential in Argentina has been evidenced in the past years. This provides a unique opportunity for Argentina to assume the role of a key player in the global energy transition race. Still, many existing challenges will have to be addressed as well as new ones which will unfold as this transition keeps evolving.

China is a big player in the lithium sector and Latin America is one of the main targets for investment by Chinese companies. In recent years, and despite the impact of the global covid-19 pandemic, a significant investment trend has occurred, especially in relation to Argentine lithium projects.

Best practices and improved regulation addressing the needs of end-users and investing heavily in lithium projects will play a crucial role. Further, capturing ESG standards is an unavoidable obligation for policy makers and private investors in order to foster the energy transition in a sustainable and coherent manner, whereby all stakeholders should be taken into account.

9

Energy Transition in the Peruvian Mining Sector: Regulatory Approach or Just Private Self-Regulation?

Enrique Velarde

9.1 Introduction

Nowadays, the world recognises energy and mining sectors as essential economic activities which provide benefits such as potential economic growth, employment, local income and development, and private investment.

In that regard, Peru is a country of ancient mining tradition, which has been preserved and cultivated by leading international companies that have been working to expand the mining activities within the country. At the global and Latin American level, and according to the information from the Ministry of Energy and Mines (MINEM for its Spanish acronyms), Peru is among the top producers of a variety of metals (gold, silver, copper, lead, zinc, iron, tin, molybdenum, tellurium, among

E. Velarde (✉)
Law Department, Pontificia Universidad Católica del Peru, Lima, Peru
e-mail: enriquevm81@gmail.com

G. Wood and J. F. Neira-Castro, *From Fossil Fuels to Low Carbon Energy Transition*, Energy, Climate and the Environment,
https://doi.org/10.1007/978-3-031-00299-1_9

others). This scenario reflects the abundance of resources, the production capacity of the Peruvian mining industry and the stability of economic policies in the country. All of which turn Peru into one of the main destinations for foreign mining project investors interested in exploration and exploitation.[1]

On the other hand, the energy sector is going through a transitional stage towards more renewable energies. However, energy transition is only achievable by strategic, and coherent legislation that facilitates responsible investment and exploitation of resources. Consequently, this document's efforts focus on describing the energy policy as in the regulatory developments in Peru, which will give the reader a broad panorama of how the mining sector and energy transition are regulated.

9.2 The Mining Industry and the Energy Transition Regulatory Framework in Peru

9.2.1 The Mining Sector in Peru

Peru, especially the Andean region, has enormous geological potential, constituting the primary source of mineral deposits in Latin America and the world. In 2020, Peru was the world's second largest producer (after Chile) of primary copper and silver, the third largest zinc producer, the world's fourth largest tin and lead producer, and the world's eighth largest gold producer.[2]

There is currently a high demand for Peruvian produced minerals in the global market, whose development is based on production and

[1] Ministerio de Energía y Minas, *Anuario Minero 2020* (Primera, 2020), https://cutt.ly/PmV2bhC; INEI, 'Instituto Nacional de Estadística e Informática INEI' (2021), http://www.inei.gob.pe/estadisticas/indice-tematico/economia/, accessed 8 October 2021; Carlos et al. Monge, *Transiciones: Post Extractivismo y Alternativas Al Extractivismo En El Perú* (Alejandra; Alayza and Eduardo Gudynas eds, Primera Ed, Centro Peruano de Estudios Sociales - CEPES 2011).

[2] Ministerio de Energía y Minas (n 1); U.S. Geological Survey, *Mineral Commodity Summaries 2021: Sand and Gravel (Industrial)* (2021).

industry. According to the MINEM, the United States, China, Switzerland, Japan, Canada, and the European Union are the main users of these minerals.[3] Peru also has great potential in non-metallic minerals, in fact, Peru is among the few countries where deposits of those minerals are found.[4]

Therefore, mineral exportation represents a significant income for the Peruvian government and is an important contributor to the national GDP. According to the Central Bank of Peru (BCRP for its Spanish acronyms),[5] in the year 2020, the mining industry's contribution to GDP was around US$ 42,663 million, representing approximately 21.12%.

Moreover, according to the Custom and Tax Authority (SUNAT for its Spanish acronyms) and the BCRP, mining exports have an important impact on Peru's economy. In 2020, they represented 60.8% of all national exportations.[6,7]

9.2.2 Mineral Policy in Peru

The 1990s represented a decade of significant social, economic, and political change in the country´s direction. Different industries, such as the mining sector, were subject to the incorporation of new regulatory and legal frameworks. Since then, the government's approach toward the mining industry has been more focused on promoting, regulating, and inspecting the sector.

Therefore, the Congress of the Republic of Peru approved, among others, the new General Mining Law in 1992 (GML) and complementary regulation such as the Environmental Code, and the Juridical

---

[3] Ministerio de Energía y Minas (n 1).

[4] Ibid.

[5] BCRP, 'SISTEMA DE CONSULTAS DE ESTADÍSTICAS' (2021), https://estadisticas.bcrp.gob.pe/estadisticas/series/anuales/resultados/PM04991AA-PM04973AA-PM05132AA-PM05133AA-PM05134AA-PM05135AA-PM05136AA-PM05137AA-PM05138AA-PM05139AA-PM05140AA-PM05145AA/html/2010/2021/, accessed 27 July 2021.

[6] SUNAT, 'Nota Tributaria - Búsqueda Por Actividad Económica' (2021), https://www.sunat.gob.pe/estadisticasestudios/exportaciones.html, accessed 28 July 2021; BCRP (n 5).

[7] Ministerio de Energía y Minas (n 1).

Stability Agreements regulation in 1992.[8] Since these changes were implemented, the governmental approach and strategy have not changed substantially, so the country remains as a mainly raw material exporter with undeveloped local industry (i.e. metallurgy, iron, and steel products). Subsequent governments have promoted investments in extractive industries in an attempt to turn Peru into a very competitive market and destination from the other mining countries.

However, the most significant changes were performed in the environmental field (i.e. the General Environmental Law in 2005) and the improvement of standards and control on industry.[9] Later on, these changes brought new governmental institutions such as the Agency for Environmental Assessment and Enforcement (OEFA for its Spanish acronyms) and the Agency of Environmental Certification for Sustainable Investment (SENACE for its Spanish acronyms).

The National Development Strategic Plan—Bicentenary Plan (Plan Estrategico de Desarrollo Nacional—El Plan Bicentenario) was approved in 2011 and was updated in 2016. These documents explain the aspects in need of consideration in order to reach sustainable development. However, they have not described in any detail how to perform such ideas and objectives. Although the documents explain the economic benefits of the Peruvian mining industry, nothing therein is written regarding other equally important aspects related to the mining sector such as industrial diversification and development that could add additional value to the minerals (i.e. refineries, metalworking, mining clusters, among others).[10]

Moreover, the MINEM has not developed any policy documentation or any political plan that would diversify the industry, reduces the dependency on raw material exports, and reach the long-desired developmental goals. There are no in-depth discussions on the long-term process,

---

[8] República del Perú, Texto Único Ordenado de la Ley General de Minería 1992 80; República del Perú, Ley General del Ambiente 2005; República del Perú, LEY DE PROMOCION DE LAS INVERSIONES EXTRANJERAS DECRETO LEGISLATIVO 1991 6.

[9] República del Perú Ley General del Ambiente (n 8).

[10] CEPLAN, 'Plan Estratégico de Desarrollo Nacional Actualizado' (2016), www.ceplan.gob.pe,
.

the future international need for minerals, or the global context (i.e. COVID-19, climate change, China's need for minerals).[11]

9.2.3 Regulatory Mining Framework

According to the Peruvian constitution, all natural resources, renewable or not, including all minerals, belong to the state. The concession system is the mechanism devised under Peruvian law to grant rights to conduct mining activities (i.e. exploration, exploitation, beneficiation plants, mineral transportation). Moreover, the mining concession (i.e. exploration and exploitation activities) is a different and independent right from the ownership, or any other surface right, over the land under which a mining concession is located.[12]

Mining concessions in Peru are granted under the grid-based system and are single concessions for exploration and exploitation. In both scenarios, no overlap between mining concessions is allowed. Therefore, titleholders have unlimited rights to explore and exploit minerals they find within their concessions (delimited using the Universal Transverse Mercator (UTM) coordinates).[13]

However, the exercise of rights derived from a mineral concession (e.g. the right to explore, develop, and exploit within the internal boundaries of the mineral concession) is subject to the awarding of required permits, authorisations, and approvals, such as the approval of the Environmental Management Instrument (EMI) designed for distinct activities such as mineral exploration or exploitation.[14]

[11] Naciones Unidas, 'Minería Para Un Futuro Bajo En Carbono: Oportunidades y Desafíos Para El Desarrollo Sostenible' (2019).

[12] República del Perú, Constitucion Politica del Peru 1993 1; República del Perú Texto Único Ordenado de la Ley General de Minería (n 8); Martin Belaunde Moreyra, *Derecho Minero y Concesion* (Editorial San Marcos E.I.R.L. ed, Cuarta, 2011).

[13] República del Perú Texto Único Ordenado de la Ley General de Minería (n 8); Belaunde Moreyra (n 12).

[14] República del Perú Texto Único Ordenado de la Ley General de Minería (n 8); República del Perú, Sistema Nacional de Evaluación de Impacto Ambiental 2001; República del Perú, Reglamento del Sistema Nacional de Evaluacion de Impacto Ambiental 2009 153; República del Perú, Reglamento Ambiental para las Actividades de Exploracion Minera 1998; República del Perú Ley General del Ambiente (n 8).

Additionally, once the mining project is finished, titleholders must reclaim or close all mining works and facilities according to a Mine Closure Plan approved by the MINEM. This plan is submitted together with the EMI. It is important to premise this by recognising that forfeiture or expiration of a mineral concession does not release or discharge a titleholder or lessee from their mine closure obligations. In fact, there are some financial mechanisms in place to fulfil this obligation (i.e. letter of guarantee or surety bond, guarantee trust, and joint and several guarantees).[15]

Complementary to this, the state conducts a consultation process and under the ILO Convention 169, when the rights of indigenous or native peoples may be directly affected by a regulatory or administrative measure. This procedure is getting very critical in Peru, considering that Peru has an important number of native and indigenous communities, mainly located in the Andeans, where most mining project development is taking place. The aim of this regulation is—in theory—to reach an agreement between the governmental entity (e.g. MINEM) and the indigenous people with regard to the implementation of regulatory or administrative measures (e.g. EMI or its modification). However, many social conflicts have arisen due to the lack of, or poor approaches in its implementation.[16]

9.2.4 Energy Transition Framework

During the last twenty years, the energy sector in Peru has registered a noteworthy growth due to the increase in internal demand, which is related to economic development. This economic boom period in Peru has seen the government developing different regulatory mechanisms

---

15 República del Perú Ley General del Ambiente (n 8); República del Perú, Ley que Regula el Cierre de Minas 2003.

16 República del Perú, Ley del Derecho a la Consulta Previa a los Pueblos Indígenas u Originarios, reconocido en el Convenio 169 de la Organización Internacional del Trabajo (OIT) 2011; República del Perú, Reglamento de la Ley del Derecho a la Consulta Previa a los Pueblos Indígenas u Originarios, reconocido en el Convenio 169 de la Organización Internacional del Trabajo (OIT) 2012; Monge (n 1); La Ley and others, '¿ Con Quién Estás de Acuerdo ? ¿ Con Quién Estás de Acuerdo ? Las Instituciones Socioambientales'.

for these monopolistic activities (i.e. energy transport and distribution services), allowing the support of a secure energy supply due mainly to natural gas (i.e. Camisea project). Consequently, the supply for other energy sources registered a slight decrease (i.e. crude oil production).[17]

On May 31, 2010, the National Energy Policy Project of Peru 2010–2040 was published and circulated to receive any contribution from the private and public sectors. Later, on November 24, 2010, by Supreme Decree N° 064-2010-EM, the government approved the National Energetic Policy of Peru 2010–2040.[18]

In general terms, this document is a long-term plan,[19] centred on sustainable development, with an emphasis on promoting and protecting the private investment, minimising social and environmental impacts, and respecting and encouraging energy markets. Moreover, the plan aims to promote energy efficiency, and the development of renewable energies at the local, regional, and national levels.[20]

Four years later, on November 28, 2014, the Peruvian Ministry of Energy and Mines presented a new proposal for the National Energy Plan (2014–2025) and sought contributions about the role of the energy sector in national development. Also, part of the objective is aligned with those established by the OECD, and meets its requirements, with an environmental emphasis.[21]

According to this plan, there are nine main objectives. These range from developing a diversified energy matrix with an emphasis on renewable sources and energy efficiency to integrating with the energy markets

[17] Petroleo y Energia (SNMPE) Sociedad Nacional de Mineria, *El Sector Minero Energetico y El Cambio Climatico* (2014); MINEM, 'Plan Energetico Nacional 2014–2025' [2014] Ministerio de Energía y Minas 9, http://www.archivogeneral.gov.co/sites/all/themes/nevia/PDF/Transparencia/ACUERDO_02_DE_2014.pdf.

[18] Republica del Peru, Decreto Supremo N° 064-2010-EM 2010; MINEM (n 17).

[19] The plan is based on the guidelines of the Strategic Plan for National Development—Plan Peru 2021, prepared by the CEPLAN.

[20] Republica del Peru Decreto Supremo N° 064-2010-EM (n 18); MEM, 'National Energy Plan 2014–2025 | Peru' 36.

[21] Republica del Peru Decreto Supremo N° 064-2010-EM (n 18); MINEM, 'Plan Energético Nacional 2014–2025' (2014); Sociedad Nacional de Mineria (n 17).

of the region, encouraging long-term commitments and market integration, through to incentives to develop the required infrastructure.[22]

It is also important to take note of what has been mentioned regarding the regulatory framework intended to be developed. This, in general terms, seeks to promote private investment in energy activities, corresponding to the state continuing the "business as usual" approach.[23]

Regarding the energy transition towards more environmentally favourable alternatives, the aim of which is the development and use of clean energies and technologies with low polluting emissions, this could be accomplished by implementing measures that mitigate emissions from energy activities, for example, promoting energy projects that obtain the benefits of emission reduction certificates for the carbon market.[24]

Moreover, the plan proposes, among other things, the substitution of liquid fuels derived from oil with natural gas and liquefied petroleum gas, the use of solid and liquid waste for energy production, and social responsibility practices in energy activities and facilitate a decentralised system in the distribution of natural gas for all sectors in Peru.[25]

The plan resulted from a diagnosis made about the situation in the Peruvian energy sector, where it was determined that emissions in Peru came mainly from: deforestation and the loss of forest biomass thereby releasing CO_2 into the atmosphere (i.e. the Amazon); the consumption of fossil energy (oil, gas and coal) for transportation, electricity generation, and fuel consumption; gases released in agricultural and livestock activities; the decomposition of solid waste and wastewater. Consequently, activities where the focus is on gross domestic product

[22] Republica del Peru Decreto Supremo N° 064-2010-EM (n 18); MINEM (n 17).

[23] Republica del Peru Decreto Supremo N° 064-2010-EM (n 18); MINEM (n 17); Patricia Patrón Álvarez, *¿Energía o REDD? Primero EAE Evaluación Ambiental Estratégica En El Perú: Próximos Pasos* (Ambiente y Recursos Naturales Derecho ed, Primera, 2011), http://www.dar.org.pe/prensa_publicaciones.htm; Sociedad Nacional de Mineria (n 17).

[24] Republica del Peru Decreto Supremo N° 064-2010-EM (n 18); Patrón Álvarez (n 23); Sociedad Nacional de Mineria (n 17); MINEM (n 17).

[25] Republica del Peru Decreto Supremo N° 064-2010-EM (n 18); MINEM (n 21); Sociedad Nacional de Mineria (n 17).

growth (i.e. agriculture, livestock, construction, mineral production) are also responsible for the increase in greenhouse gas emissions.[26]

The objective was to make improvements aimed at achieving a sustainable, safe, reliable, timely, and accessible energy supply for all sectors in Peru, with an emphasis on climate change. This is a problem that Peru must face by reducing the contribution of liquid and gaseous hydrocarbons in the energy matrix and look for renewable energy alternatives (such as hydroelectricity).[27]

9.3 Energy Transition in the Mining Industry

The mining industry in Peru fits within the definition of predatory extractivism, characterised by, among other aspects, its aggressive expansion within its different stages, environmental liabilities, violation of human rights (i.e. indigenous communities), forced displacements, and conflicts. As a result, this industry is faced with many social conflicts, showing serious governance problems and a lack of an adequate institutional framework allowing companies and the government the capacity or tools to address conflicts and turn them into opportunities.[28]

Moreover, this industry is energy intensive with permanent access to a stable electricity source. With every new mining project, the demand for energy increases, especially due to the implementation of new technologies and the automation and electrification of mining sites.[29,30]

Currently, the electricity produced and acquired by mining companies is based primarily on fossil fuels (i.e. Camisea natural gas project) and hydropower plants. However, despite the declining of costs related to other renewable energies (i.e. solar and wind), neither the mining

[26] Republica del Peru Decreto Supremo N° 064-2010-EM (n 18); MINEM (n 17); Sociedad Nacional de Mineria (n 17).

[27] Republica del Peru Decreto Supremo N° 064-2010-EM (n 18); MINEM (n 17); Sociedad Nacional de Mineria (n 17).

[28] Monge (n 1).

[29] Nicolas Maennling and Perrine Toledano, 'La Energía Renovable En La Minería: Acelarando La Integración de Energías Renovables' (2018), https://bit.ly/2yv8OZP.

[30] MINEM (n 17).

industry nor the Peruvian government has established any actions that would contribute to preventing global temperature rise.[31]

However, for a developing country like Peru that relies on models which in turn depend on an economy based on raw material exportation, the transition towards a more sustainable economy is a challenge. This requires political willingness and institutional efforts to implement coherent public policies, programmes, and plans.[32,33]

According to Monge, the transition to sensible mining practices implies breaking the predominance of private self-regulatory mechanisms, which are the result of a defensive and intransigent mining industry swaying the legal framework. The proposals are linked to create self-regulatory mechanisms (i.e. codes of conduct and corporate social responsibility practices), including those related to the energy transition.[34]

These mechanisms recognise the existence of certain negative externalities, a consequence of their activities, and that mining companies have the capacity to control them without the need to create new regulation.[35] The efforts related to climate change from the mining industry focus on aspects linked to the stakeholders more than the internal procedures that would help in reducing energy consumption or performing an energy transition.[36]

Nevertheless, the National Energy Plan aims to achieve low carbon emissions by promoting the use of clean energies, with broad or vague objectives, and without indicators or targets with which to measure the progress and results. Moreover, the plan indicates that in the next period from 2014–2025, new mining operations are the main factor that will increase final energy consumption.[37]

---

31 Maennling and Toledano (n 29); Monge (n 1).

32 Maennling and Toledano (n 29).

33 International Energy Agency (IEA), 'The Role of Critical Minerals in Clean Energy Transitions' [2021] IEA Publications.

34 Monge (n 1).

35 ibid.

36 Sociedad Nacional de Mineria (n 17).

37 MINEM (n 17).

However, the National Energy Plan does not reflect an important intersectoral coordination, lack of consistency in public policies, and little progress on the diversification of the Peruvian energy matrix. As a matter of fact, there are no specific or clear actions that link energy transition and the mining industry, such as incentives for the transition and use of renewable energies. This reflects the Peruvian government's position to keep the "business as usual" approach instead of, for instance, adopting an energy governance criteria.[38]

An example is the prioritisation of natural gas exportation over the impulse of industrialisation to satisfy national demand and avoid a supply crisis due to continuous growth. Peru has a high demand for fuels, especially natural gas, and this scenario generates concern about the future availability of natural gas for the domestic market and its 'possible' industrialisation.[39,40,41]

However, the promotion of hydroelectric plants in the Peruvian Amazon (i.e. the energy agreement between Peru and Brazil) does not represent a solution or option because the implementation of these projects will result in deforestation, involuntary displacement of local populations and indigenous peoples, as well as possibly producing methane, contributing further to the generation of greenhouse gases.[42,43,44]

Finally, there is no strategic planning from the government, an approach which makes it difficult to identify Peruvian demand and reserves for more efficient, sustainable, and equitable use of energy. Also, the National Energy Plan does not show any guidelines for a gradual change of the energy matrix, ensuring a balance between sources and use.[45]

---

[38] Monge (n 1); Ivan Lanegra, *El Camino Ambiental Hacia La OCDE* (2018); MINEM (n 17).

[39] Macroconsult, 'Impacto Económico de La Minería En El Perú' (2012).

[40] Naciones Unidas (n 11).

[41] International Energy Agency (IEA) (n 33).

[42] Patrón Álvarez (n 23).

[43] Vanessa Cueto La Rosa, *Buscando La Gobernanza Energética En El Perú* (Primera, 2011).

[44] Denise Humphreys Bebbington and others, 'Evaluación y Alcance de La Industria Extractiva y La Infraestructura En Relación Con La Deforestación: Amazonía' 76.

[45] Monge (n 1).

9.4 Conclusions

Peru has enormous geological potential in the Andeans, constituting the primary source of mineral deposits in Latin America and the world that at the same time has represented a considerable income (i.e. national GDP) for the Peruvian government. However, during the last twenty years, the energy sector in Peru has registered remarkable growth due to the increase in internal demand. The Peruvian government has published a plan centred on sustainable development, emphasising the promotion and protection of private investment, minimising social and environmental impacts, and respecting and encouraging the energy market.

The National Energetic Policy of Peru 2010–2040 has concluded that the origin of the emissions is: (i) loss of forest biomass that releases CO_2 into the atmosphere due to deforestation (i.e. the Amazon); (ii) the consumption of fossil energy (oil, gas, and coal) for transportation, electricity generation and fuel consumption; (iii) gases released in agricultural and livestock activities; and (iv) the decomposition of solid waste and wastewater.

Although the objective of this chapter was to improve the way to achieve a sustainable, safe, reliable, timely, and accessible energy supply for all sectors in Peru (emphasising climate change), there is no evidence that the mining industry is moving or contributing to reducing the use of liquid and gaseous hydrocarbons in the energy matrix, nor is it apparently looking for alternatives in the form of renewable energies. Moreover, the mining industry has an intensive use of energy and permanent access to a stable electricity source. Still, despite the declining costs related to other renewable energies (i.e. solar and wind), neither the mining industry nor the Peruvian government has established actions in the National Energy Plan to prevent the rise of global temperatures, with the implementation of these energies on their mining sites.

The mining industry's efforts related to climate change focus on aspects linked to the stakeholders more than the internal procedures that would help reduce energy consumption or play a role in the energy transition. The transition to a sensible mining activity implies breaking the predominance of private self-regulation mechanisms, which would change the current legal framework, something the mining industry is not looking for.

Moreover, the National Energy Plan does not reflect crucial intersectoral coordination but a lack of consistency on public policies and little progress on the diversification of the Peruvian energy matrix. Consequently, no specific or clear actions link the energy transition and the mining industry, reflecting that the Peruvian government keeps the "business as usual" approach.

The National Energy Plan does not show any guidelines for a gradual change of the energy matrix, ensuring a balance between sources and use, reflecting the lack of government planning, which makes it difficult to identify the Peruvian demand and reserves for more efficient, sustainable, and equitable energy use.

Part III

Social License to Operate and Dispute Resolution Mechanisms

10

Social Licensing to Operate and Social Justice in Energy Transition: The Case of Brazil

Thaiz da Silva Vescovi Chedid, Eduardo Guedes Pereira, Edmilson Moutinho dos Santos, and Hirdan Katarina de Medeiros Costa

10.1 Introduction

This work intends to verify aspects of social license to operate (SLO) and social justice in the energy transition. The need and acceptance of the people to any energy project are key in any part of the world, but especially in Latin America. This chapter's main objective is to describe how the concept of social justice, consultation, and benefits for

T. da Silva Vescovi Chedid (✉) · E. M. dos Santos ·
H. K. de Medeiros Costa
Institute of Energy and Environment, University of São Paulo, São Paulo, Brazil
e-mail: thaizvescovi@hotmail.com

E. M. dos Santos
e-mail: edsantos@iee.usp.br

H. K. de Medeiros Costa
e-mail: hirdan@usp.br

G. Wood and J. F. Neira-Castro, *From Fossil Fuels to Low Carbon Energy Transition*, Energy, Climate and the Environment,
https://doi.org/10.1007/978-3-031-00299-1_10

the community address the needs of the people to allow the energy transition and what concerns the interest in social and environmental public policies inserted in the Brazilian oil and gas sector. This will be the theme of this chapter.

The energy industry is one of the most relevant in Brazil, being responsible for a large generation of foreign exchange and economic development.[1] However, this economic activity is associated with significant environmental and social impacts. The guiding question of this chapter is to understand how the Brazilian government can combine the need to protect communities and the environment while developing its energy industry during the energy transition.

It seeks to address the issue from a case study of Brazilian energy and an overview of the relevant law to subsequently bring some specific content such as SLO and its requirements, the socio-environmental context, and the challenges involved. After that, social justice in the Brazilian energy sector, socio-environmental protection, mitigating measures, and public consultations in accordance with Brazilian legislation will be discussed.

This work uses a deductive and exploratory research method justified by empirical analysis, namely the methodological technique of bibliographic and legal analysis. From this, inductive methods were used, according to scientific convenience. First, the work seeks to introduce the reader to the topic by presenting a background to the Brazilian energy sector. The first section of this chapter then focuses on social licensing to operate (SLO) highlighting the social environmental context and the challenges involved. Therefore, it takes care to present the importance of social justice in the Brazilian energy sector, including socio-environmental protection, mitigation measures, and public consultation in accordance with Brazilian law. In developing the case, possible

E. G. Pereira
Natural Resources and Energy Law, Siberian Federal University, Krasnoyarsk, Russia

[1] LEME, Alessandro Andre. A reforma do setor elétrico no Brasil, Argentina e México: contrastes e perspectivas em debate. Available at: https://www.scielo.br/j/rsocp/a/M9Xq9hqmMnFDfdTCQCQPQ8N/?lang=pt. Accessed june 2021.

scenarios are brought to light between the future of the theme and the lessons to be learned by Brazil.

Finally, the theme shows relevance because it deals with perceptions of the community and its acceptability between a company and its local operations. This kind of informal authorization provided by society to enterprise has been shown to be important. Such arrangement has been shown to be win–win situations: the local population is happier when they feel they have a say in company actions and decisions, and the company gains greater trust and acceptance for its activities.

10.2 The Brazilian Energy Sector

10.2.1 Background

The Brazilian energy sector is guided by a set of objectives that aim to: guarantee access to a quality service at a fair price for the entire population, maintaining strict commitments to preservation of the environment, and the sustainable management of natural resources. Such policy simultaneously contributes to economic and social progress while maintaining one of the cleanest energy sources in the world.[2]

The concern with an external dependence on fossil fuels has led to greater diversification of energy sources, with a preference for renewable sources with low environmental impact. Today, Brazil is internationally recognized for its pioneering spirit in the development of efficient and environmentally sustainable energy alternatives, in particular ethanol.[3]

In this context, it is important to highlight the present energy transition: the transition to a lower carbon economy was already a trend when the COVID-19 pandemic emerged. As a consequence, it can be

---

[2] Ibid.

[3] Manuilova, Anastassia; Suebsiri, Jitsopa; Wilson, Malcolm. Should Life Cycle Assessment be part of the Environmental Impact Assessment? Case Study. 2009.

seen that behavioral changes can be accelerated. Therefore, large companies in the sector anticipated strategy adjustments and sought to become carbon neutral by 2050.[4]

The planning studies developed by EPE (Empresa de Pesquisa Energética) incorporate the concept of socio-environmental sustainability. The focus here is on more sustainable alternatives based on current international discussions and negotiations on climate change, as evidenced at COP-15, held in Copenhagen, and reiterated at the Cancun Conference (COP-16). Commitments made by Brazil in these international negotiations were formalized through Law n.12.187/09, establishing a voluntary target for the mitigation of greenhouse gas emissions in 2020, shifting discussions on climate change in Brazil to new institutional level.[5]

In this context, the PDE 2020 presents itself as an important instrument in creating a mitigation plan, as it incorporates measures that, together, contribute to the reduction of GHG emissions in energy production and consumption. These include increased energy efficiency, increased installed hydroelectricity parks, alternative sources of electrical energy such as wind, biomass, and small hydroelectric plants (SHP). In addition, the evaluation of sugarcane expansion areas needed to increase the biofuel volume and the consequent replacement of fossil fuels are among the measures incorporated allowing the country to maintain a similar level of emission intensity as verified in 2005.[6]

The 1990s were marked by important changes in the way the Brazilian energy sector worked. As with all infrastructure sectors, the responsibility for energy investments was assigned to the State, either for strategic or for economic reasons. From the 1980s onwards, the exhaustion of the replacement model of imports, the fiscal crisis, and the consequent inability of the State in providing the necessary resources to the provision

[4] Ibid.

[5] EPE. Empresa de Pesquisa Energética. Ten-year energy plan 2029. Brasília, 2019. Available at https://www.epe.gov.br/pt. Accessed on 06/12/2020.

[6] EPE. Empresa de Pesquisa Energética. Indicative transportation pipeline plan. Brasília, 2019b. Available at https://www.epe.gov.br/pt. Accessed on 06/12/2020.

of inputs and services, generated important changes in various sectors of the Brazilian economy.[7]

Due to political impasses and the lack of a social consensus on the best alternative to the particular model of economic growth pursued, Brazil failed to overcome the crisis experienced by the country. As a result the 80's are now referred to as the "lost decade." In the 1990s, government initiatives turned to the promotion of a new operating model for the economy.[8] Taking advantage of the favorable surge in external capital resources, economic policies removed barriers to trade and foreign investments. These changes combined with a successful price stabilization plan initiated reform of the infrastructure sectors. This was all made in the context of international financial turmoil and the permanent need to resolve structural problems such as public debt.[9]

Brazilian energy sector reforms, as well as other infrastructure sectors, contain multiple objectives. Among these are the reduction of public debt and the creation of competitive structures when possible. As a result, there was a need to establish new regulatory frameworks, representing a historic institutional transformation in Brazil, and create regulatory agencies which would become the way of decentralizing the role of the state in these sectors.[10]

In Brazil, the constitution of regulatory agencies represents a milestone in its regulatory tradition, in light of the tradition of implicit regulation of infrastructure companies, or by the National Department of Water and Electric Energy, or by the National Department of Fuels, bodies that are subordinate to the Ministry of Mines and Energy.

---

[7] Manuilova (n 3).

[8] Rathmann, Régis Rathmann. Clean Energy Technologies Analysis Using RETScreen. Brasília: MCTIC, 2017.

[9] Ibid.

[10] Costa, Hirdan Katarina de Medeiros; Mussarra, R. M. L. M.; Miranda, M. F.; Moutinho dos Santos, E. Environmental License for Carbon Capture and Storage (CCUS) Projects in Brazil. Journal of Public Administration and Governance, v. 8, p. 163–185, 2018.

In this traditional configuration, sectorial policies were directly subordinated to the executive branch, implying the pursuit of often contradictory goals, such as microeconomic (productive efficiency), macroeconomic (control of BNDES Tests nº 11 6 inflation and public deficit), and social (universalization of services).[11]

It should be added that during the period of state dominance, no effective form of social regulation was exercised. The reason for this is the States natural disinterest in self-inspection and that companies were under its shareholding control with no mechanisms to control them.

In this way, the constitution of the agencies enhances the emergence of new actors in the sectorial arena of technical decisions, especially with regard to a greater balancing between them. In the case of Brazil, the National Agency for Electric Energy (ANEEL) by Law 9.427, of December 26, 1996,[12] and the National Petroleum Agency (ANP) by Law 9.478, of 6 August 1997.[13]

The reforms of the Brazilian energy sector introduced the independent agencies (ANEEL and ANP), which represented an important innovation from the point of view of the organization of the State.[14] For a country with the political tradition of Brazil, both these agencies could contribute to implementation of guidelines and help overcome inherent difficulties therein. Furthermore, gains for citizenship could also be brought through greater transparency in carrying out the regulatory mission.

---

[11] Ibid.

[12] Governo Do Estado. Law No. 9427 of December 26th, 1996. Available at: http://www.planalto.gov.br/ccivil_03/leis/l9427compilada.htm. Accessed 14 june 2021.

[13] _____. Governo Do Estado. Law No. 9478 of August 6th, 1997. Available at: http://www.planalto.gov.br/ccivil_03/leis/l9478.htm. Accessed 14 june 2021.

[14] Aneel. Agência Nacional de Energia Elétrica. Normative Resolution n°583 of 2013. Estabelece os procedimentos e condições para obtenção e manutenção da situação operacional e definição de potência instalada e líquida de empreendimento de geração de energia elétrica. Brasília, 2013.

10.2.2 Social License to Operate (SLO)

The greater inclusion of interested parties widens the variety of actors in the process of corporate governance. The resolution of conflicts relating to environmental and social impacts is understood to be the responsibility of the State and corporations. These demand collaborative management through agreement, consensus, and public participation in order to deliver a SLO involving civil society.[15]

Large-scale natural resource extraction projects (including exploration and processing activities) profoundly transform environments, communities, and economies, and often generate social conflict, e.g., regarding the relationship between the environmental and social risks experienced and interpreted by local communities, and the business risks experienced and interpreted by corporations. Findings reveal that, at least in the case of the extractive industries, these two types of risk together can co-constitute a company–community conflict and its role in the regulation of sustainability performance in the extractive industries.[16]

For the energy sector in particular, this argument is justified, because, despite its importance in regards industrial activity, the sector is subject to societal criticism due to the negative impacts brought about by its operations. Therefore, the decision-making process in this sector needs to seek a balance between progress and productivity, pursue a social approach valuing dialogue while absorbing different points of view and collective interests.[17]

The SLO can be understood as a conceptual approach applied by organization when engaging with communities and stakeholders to assist in the identification of proactive actions. Therefore, this company approach is taken to increase the likelihood that they will enjoy the support of the communities where they work.

SLO proposes a systemic vision, integrating social challenges along with the usual resource technological, production, and management

---

[15] Franks et al. (2014). Conflict translates environmental and social risk into business costs. Proceedings of the National Academy of Sciences, 111(21), 7576–7581.

[16] Ibid.

[17] Franks et al. (n 15).

challenges and is directly linked to the way companies define and understand sustainable development.[18] Thus, this kind of informal authorization provided by society to companies has been shown to be important, since both win: interested parties and the affected population benefit when they feel they have a stake in the decision-making process, and the company gains in terms of security and acceptance allowing undertaking of and completion of activities.[19]

For an enterprise to obtain a SLO, it is necessary that the project in question has ensured legal, social, and economic legitimacy.[20] In order to support the discussions here, SLO research presents guidelines on how to establish good communication at all stages of project implementation with the parties involved and managing conflict resolution.[21]

Thus, the social license has attracted the attention of companies that operate in sectors with the greatest socio-environmental impacts. As a result, it is a well-established narrative throughout the world with its acceptance being particularly widespread within the energy and mining sectors.[22]

For mining companies, the concept of a SLO is based on the idea that they do not need government authorization, instead a "social permit," i.e., a SLO is required to conduct business. In Brazil, SLO has been incorporated through institutional, foundational, and company rhetoric, with a set of practices associated with it.[23]

A SLO becomes part of a company's business strategy and planning and development of productive operations. It is understood as necessary for the reduction of business risks associated with resistance and conflicts within society. Such consequences can directly affect the company's

[18] Bice, Sarah; Moffat, Kieren. Social licence to operate and impact assessment. Available at: https://www.tandfonline.com/doi/full/10.1080/14615517.2014.950122. Accessed May 2021.

[19] Ranagen, Christopher. The emergence of social licence necessitates reforms in environmental regulation. Available at: http://www.ecologyandsociety.org/vol23/iss3/art24/. Accessed May 2021.

[20] Prno, Jason, Slocombe, Scott. A Systems-Based Conceptual Framework for Assessing the Determinants of a Social License to Operate in the Mining Industry. Available at: https://link.springer.com/article/10.1007%2Fs00267-013-0221-7. Accessed May 2021.

[21] Ranagen (n 19).

[22] Franks et al. (n 15).

[23] Ranagen (n 19).

profitability, production, and reputation which in turn can lead to repercussions in communication spheres leading to increased levels of government regulation.[24] This has resulted in an increased perception by companies of the costs and risk posed to businesses associated with conflicts between enterprises and local communities.[25]

Moving toward a typology of conflicts and costs shows that the main conflicts occur due to issues such as pollution, distribution of wealth, access or competition for resources, consultation and communication with the community, health and security of the community, culture, and customs. In regard to costs, these are mainly due to modifications or abandonment of projects, consultancy costs to manage reputation, difficulties in raising capital, and time dedicated to managing conflicts and fines.[26]

Companies are increasingly aware of the risks and costs to business associated with conflicts with local communities. From the company's point of view, a SLO contributes by minimizing risk to business and guaranteeing reputation.[27]

In this context, the SLO adds to the risk management agenda, a scenario where a SLO considers social risk. In summary, the SLO reflects the understanding that the compliance of legal regulations in isolation is, for the most part, incapable of resolving all the society's expectations. The SLO corresponds to a necessary, but intangible, part of the contract that exists between a company and the company or social group, which allows and maintains business operations.[28]

In the energy industry, there was a need to respond to technological and management challenges but also incitement and social pressure. According to Franks and Cohen, risk mapping tends to deal with neutralizing technological barriers and separating their research projects from social influences. SLO collaborates to fill this gap by proposing a

---

[24] Parson et al. American journal of international law. Cambridge University Press:2014.

[25] Franks et al. (n 15).

[26] Owen, John R., Kemp, Deanna. Social licence and mining: A critical perspective. Available at: https://www.sciencedirect.com/science/article/abs/pii/S0301420712000529. Accessed 10 May 2010.

[27] Ibid.

[28] Franks et al. (n 15).

systemic approach that integrates these two areas (social and technology, production, and management).[29]

The granting of a SLO can be considered a kind of legitimization of the company by its stakeholders. The reality of each case and its social construction must be analyzed.[30,31] This should be in addition to governance perspectives in the scope of the business and the conditions of uncertainty present in this legitimation process.[32]

The two concepts of "reputation" and "social license" are closely related and enjoy a good corporate reputation, with local communities helping to strengthen conditions of social license and thereby reducing the risk of local conflict and opposition. SLO researchers have proposed strategies for obtaining and maintaining SLOs, including the need for clear communication before and after the implementation of the project, transparency in the dissemination of information and the development of mechanisms for conflict resolution and for processes of culturally appropriate decision-making.[33]

The multidimensionality of relationships for obtaining a license depends on the quality of relationships established, to allow a better observation of the relationships and interests and how they are developed over time, as their change can compromise the process (Brueckner et al., 2014).[34]

---

29 Franks et al. (n 15).

30 Berger, Peter, Luckmann, Thomas. A construção social da realidade: tratado de sociologia do conhecimento. Rio de Janeiro: Vozes, 2011.

31 Berger, Robert. Using the Social License Concept to Appreciate Conflict in Stakeholder Networks. Available at: https://www.researchgate.net/publication/291358622_Using_thc_Social_License_Concept_to_Appreciate_Conflict_in_Stakeholder_Networks. Accessed May 2021.

32 Funtowicz, Silvio, Ravetz, Jerry Ravetz. Ciência pós-normal e comunidades ampliadas de pares face aos desafios ambientais. Available at: https://www.scielo.br/j/hcsm/a/5R7X43J9DXT7TZsy8pxp3hR/abstract/?lang=pt&format=html&stop=previous. Accessed 10 May 2021.

33 Gunningham, Robert A. Kagan; Thornton, Dorothy. Social License and Environmental Protection: Why Businesses Go Beyond Compliance. Available at: https://onlinelibrary.wiley.com/doi/abs/10.1111/j.1747-4469.2004.tb00338.x. Accessed 10 May 2021.

34 Brueckner [Citation missing].

10.2.3 Socio-Environmental Context

Regarding the socio-environmental context, some authors claim that concerns over environmental protection influence the granting of SLO by communities, therefore, making it necessary for a company to maintain a clear dialogue on mitigation and compensation measures regarding possible impacts.[35]

In Australia, social issues tend to not receive as much attention in the arena of sustainable development as does environmental and employment when companies articulate the SLO. However, for a transformation to be successful considering eco-innovation, the inclusion of social factors is essential.[36]

Some studies have shown the reality of the relationship existing between the SLO and specific groups, revealing that these groups may be more sensitive to the installation and operation of a particular enterprise. Communities in extreme poverty expect companies to put it into practice actions aimed at creating jobs and improving local transport infrastructure, with the further objective of providing necessary developmental tools. This is in contrast to the vision of the poorest in society wait only for charity and assistance.[37,38,39]

From a gender perspective, there are challenges in developing a SLO aimed specifically at women. Even well-intentioned projects designed to bring development to a local community may pose a significant risk of further harming those already disadvantaged.[40]

---

[35] Bice (n 18).

[36] Provasnek, Anna Katharina Bräuer et al. Sustainable Corporate Entrepreneurship: Performance and Strategies Toward Innovation. Available at: https://www.researchgate.net/publication/280832103_Sustainable_Corporate_Entrepreneurship_Performance_and_Strategies_Toward_Innovation. Accessed May, 2021.

[37] Mcintyre, Michael L et al. Do firms seek social license to operate when stakeholders are poor? Evidence from Africa. Available at: https://www.emerald.com/insight/content/doi/10.1108/CG-01-2014-0011/full/html?fullSc=1&fullSc=&mbSc=1&fullSc=1&fullSc=1. Accessed May 2021.

[38] Musarra, Raíssa Moreira Lima Mendes; Costa, Hirdan K. De Medeiros. Comparative International Law: The Scope and Management of Public Participation Rights Related to CCS Activities. Journal of Public Administration and Governance, v. 9, p. 93–109, 2019.

[39] Costa (n 10).

[40] Vanclay, Frank et al. Community engagement and social licence to operate. Available at: https://www.tandfonline.com/doi/full/10.1080/14615517.2014.927108. Accessed May 2021.

Indigenous communities, on the other hand, represent groups conventional in the drafting of the SLO, as they have historical governance rights and distinct constitutional requirements. The impact of extractive projects, due to their scale and longevity have the potential to seriously affect the "cultural integrity of indigenous communities" and for a tangible SLO process to be effective it is necessary to establish a lasting partnership in regards the decision-making process concerning the use of resources.[41]

For traditional forest reserve communities, the implementation and operation of ventures can generate local conflict over the use and occupation of land, leading to an inevitable failure in obtaining the SLO. Another problem that mainly affects at risk groups occurs when an enterprise suddenly ends its activities after winning an SLO from the community affected.[42]

In this case, generating impacts and conflicts, non-contractual sacrifices and investments (social, psychological, economic, and environmental) represent an unaccounted cost of failure. It is worrying that, in a society increasingly defined by uncertainty and risks (environmental, social, political, economic), these issues are not considered during project planning and that adequate insurance is not provided for the communities involved in such social licenses.[43]

The processes that involve management of environmental resources need clarity and transparency, something difficult to obtain from an SLO concept still under construction and subject to different interpretations. A more formal methodology for reaching key elements of the SLO, such

---

[41] Ruckstuhl, Katharina et al. Māori and mining: Indigenous perspectives on reconceptualising and contextualising the social licence to operate. Available at: https://www.tandfonline.com/doi/full/10.1080/14615517.2014.929782. Accessed May 2021.

[42] Westoby, Peter et al. Dialogical (hermeneutical) and phenomenological approaches to [community] development practice. Available at: https://www.researchgate.net/project/Dialogical-hermeneutical-and-phenomenological-approaches-to-community-development-practice. Accessed May 2021.

[43] Browne, Alison Leigh et al. Social licences to operate: for better not for worse; for richer not for poorer? The impacts of unplanned mining closure for "fence line" residential communities. Available at: https://www.tandfonline.com/doi/abs/10.1080/13549839.2011.592183. Accessed May 2021.

as trust and legitimacy, could ensure that the path to establishing an SLO does not needs to be remodeled with each applied development.[44]

Thus, in the socio-environmental context, the role of the SLO is to bring the focus on to social conflict resolution brought about by the enterprise, and on the concern and care for the socially most vulnerable minorities, such as the poor, women, indigenous people, among many others. This is an important point to be made because nowadays environmental concerns might attract more attention than that given to social issues.[45]

10.2.4 Challenges

Preserving the environment is important in maintaining the lifestyle of communities, especially those that depend most directly on local natural resources. In this sense, one of the major challenges for developmental organizations is to ensure that the interests of all vulnerable groups are not neglected. When an organization starts working in a community, it is relatively easy to connect with people active in dominant positions and with organized community groups, but it is difficult to reach those who are silent or unable to defend their interests.[46]

Understanding that companies have responsibility and play a role in society, in the environment and local life is very important (Bickford et al., 2017). Gaining acceptance is particularly relevant for companies working in sectors where the risk to their reputation is high, as in the extractive sector, and not having an SLO can have significant consequences for the projects.[47]

However, there are still gaps between the theoretical ideals of the social license and the ways used to obtain it (BICE, 2014).[48] Caution must

[44] Van Putten et al. Conceptualizations of the societal perspective within economic evaluations: A systematic review. Cambridge University Press: 2018.

[45] Ehrnström-Fuentes, Maria, Kroger, Markus Kröger. In the shadows of social licence to operate: Untold investment grievances in latin America. Available at: https://www.sciencedirect.com/science/article/abs/pii/S0959652616314536?via%3Dihub. Accessed May 2021.

[46] Vanclay (n 40).

[47] Ibid.

[48] Bice (n 18).

be exercised before stating that a company, investment, or industry has achieved a comprehensive SLO at the local level.

The use of the term SLO is spreading to different industries, professions, and socio-political sectors, but not always with the same understanding of its connotations and implications. The danger is that the meaning become so obscure that its value as a tool for promoting collaboration becomes obscured or lost.[49]

In part, the studies analyzed sought answers to conflicts that permeate the SLO. This was done through observing the interactions between stakeholders and the license, and the practice involved in this process. The legitimacy of the knowledge and information underlying an SLO, and the result of policies that are being retained or granted are beginning to be questioned by the main stakeholders.[50]

The way the SLO concept is understood and passed from the company to the community can open space for conflicts and tarnish reputations built by the company.[51] So, it is interesting to consider that the social license could be understood as a continuum of multiple licenses obtained at various levels of society or from different specific places and times and requiring a change of behavior that realizes the importance of these issues.[52]

While a social license represents a shift in power relations, that shift ends up being limited by discursive pressures to legitimize mining operations, restrict issues of social license at the local level, minimize regulatory impositions, marginalize dissent, and manage reputation.[53]

For this reason, SLO campaigns cannot be based solely on the accumulation and presentation of scientific evidence, they must also involve

49 Ehrnström-Fuentes (n 45).

50 Cullen-Knox et al. Contemporary Challenges in Environmental Governance: Technology, governance, and the social licence. Available at: https://onlinelibrary.wiley.com/doi/abs/10.1002/eet.1743. Accessed May 2021.

51 Vanclay (n 40).

52 Alenn, Diana et al. Comparative Analysis of Hydraulic Fracturing Wastewater Practices in Unconventional Shale Development: Newspaper Coverage of Stakeholder Concerns and Social License to Operate. Available at: https://www.mdpi.com/2071-1050/8/9/912. Accessed May 2021.

53 Parsons (n 24).

contemporary expressions of environmental activism that seek regulatory intervention in corporate practices.[54]

Although the ultimate goal of obtaining an SLO is to improve social, environmental, and economic risks, there is still the apparent risk of political decisions and ineffective results. Schemes for governance must be adaptable to deal with changes in information, attitudes, values, and beliefs.[55]

Thus, one of the biggest future challenges for the SLO comprises changing a reactive or compensatory stance for one that is more contemplative, collaborative and strategic. Speed of information dissemination that facilitates community engagement, ways of disclosing more detailed impacts to attract investors, governance based on an accountability structure to assess performance and level of interaction with stakeholders and accumulation of risks are some of the changes that may be included in future scenarios and evolution of SLO.[56]

10.3 Social Justice in the Brazilian Energy Sector

10.3.1 Socio-Environmental Protection and Mitigation Measures

When it comes to socio-environmental protection, it is possible to glimpse not only the environmental impacts on the landscape and morphological changes, but also alteration to the biophysical structure affected by introducing turbines, alternating, and even destroying

---

[54] Murphy-Gregory, Hannah. Governance via persuasion: environmental NGOs and the social licence to operate. Avaible at: https://doi.org/10.1080/09644016.2017.1373429. Accessed May 2021.

[55] Cullen-Knox (n 50).

[56] Langaro et al. Viabilidade técnica, econômica e ambiental da utilização da energia eólica no Brasil e no Paraná. VI Semana de Estudos de Engenharia Ambiental. Unicentro/Denam. Available at: http://tede.unioeste.br/handle/tede/3030. Accessed May, 2021.

vegetation cover and even possibly seeing characteristic elements of the impacted landscapes disappearing.[57]

Moreover, there are environmental impacts on ecosystems affecting flora and fauna, with deforestation resulting in the loss of resting places, feeding and reproduction sites for important species that use the site of the park. Additional impacts are birds and bats colliding with wind turbines and the danger of electrocution on power lines, fish killed in water turbines, changes in the migratory route of birds, and the movement of people and vehicles causing accidents and incidents with animals within the park.[58]

In addition to the impacts already reported, there are also possible impacts on soil and water, due to oil and fuel spills, and products leaking into water courses and, where the original vegetation cover has been removed, soil erosion. Another impact which needs highlighting is that of noise pollution produced by turbines. This can cause discomfort for the neighboring population and passers-by. There is also a visual impact, which can bring real-estate devaluation on account of the giant pinwheels (Camargo, 2005).[59]

However, the most important thing is to not only point out the (dis)advantages, but to highlight concerns over mitigation and the adaptive measures implemented to reduce such impacts, especially when it comes to vulnerable areas such as the semi-arid Brazilian area. The Brazilian Law No. 12,187, of December 29, 2009, which institutes on the National Policy on Climate Change (PNMC) and other measures, discusses important concepts such as mitigation, adaptation, and vulnerability. According to this law, Article 2 states mitigation can be defined as being the "changes and replacement technologies that reduce the use

[57] Jeyaretnam, Terence. The evolving nature of social licence to operate. Available: https://search.informit.org/doi/abs/10.3316/informit.932163043675031. Accessed May 2021.

[58] Ibid.

[59] [Citation missing].

of resources and emissions per unit of production, as well as the implementation of measures that reduce greenhouse gas emissions and increase sinks."[60]

The same law in this article defines the adaptation process as being "those initiatives and measures to reduce vulnerability of natural systems and human beings in view of the current and expected effects climate change." Furthermore, it defines vulnerability as being "degree of susceptibility and incapacity of a system according to its sensitivity, adaptability and, character, magnitude and rate of change and variation of the climate to which it is exposed [...]. It aims to guarantee sustainable development, which is the condition for facing climate change whilst reconciling the meeting of common needs and particulars of populations and communities who live within the national territory."[61]

Currently, the biggest discussion is focused on sustainability of the planet, the use of renewable energy sources and zero pollution. Therefore, there has been an increase in research, projects, and investments that rely on government support and are aimed at bringing the production of sustainable energy to prominence.

In this scenario, Brazil is a country with its very own unique characteristics compared with its Latin American neighbors. Beginning with the huge extent of its vast territory, the country presents its own set of particular features such as land use issues associated with deforestation and the placement of energy infrastructures. The combination of these factors places Brazil, in terms of GHG emissions at a level similar to some developed countries. Regarding the transport sector (passengers and freight), this is performed mostly via the use of fossil fuel energy.[62]

Despite being a large emitter of CO_2, the country has in its favor an essentially clean electricity (power) matrix and has maintained that characteristic despite droughts and so on. Its environmental agenda is

---

[60] Brasil, Governo Do Estado. Law No. 12,187, of December 29th, 2009. Available at: https://www.camara.leg.br/proposicoesWeb/prop_mostrarintegra?codteor=841507&filename=LegislacaoCitada+-.Accessed 20 Jun 2021.

[61] Ibid.

[62] Koifman, Sergio. Geração e transmissão da energia elétrica: impacto sobre os povos indígenas no Brasil. Avaible at https://www.scielo.br/j/csp/a/X9cgs7g3nBn63SD3Cg85SpR/?lang=pt. Accessed July 2021.

dictated by the Ministry of the Environment (MMA), the departmental body responsible for the elaboration of the Brazilian Intended Nationally Determined Contributions (iNDC) under the Paris Agreement. The main goal presented in this document was the intention to reduce emission levels by 43% by 2030, reaching 1.2 $GtCO_2$ (GWP-100; IPCC AR5) in relation to the 2.1 $GtCO_2$ emitted in 2005.[63]

The country also created intermediary emission and efficiency goals, such as reaching 2 $GtCO_2$ of gross emissions in 2020 and make a 75% reduction in GHG emissions per unit of GDP by 2030 (in relation to 2015). One way by which Brazil intends to achieve these goals is through the increased use of renewable energy.[64]

Yet, it is important to highlight, even if briefly, the sectors activity viewed through the scope of indigenous lands. Such an example is the electricity sectors expansion in Brazil. This has historically resulted in situations that have led to direct or latent conflict with various indigenous communities. These problems occurred mainly as a result of consequences arising from the construction of hydroelectric power plants and high voltage transmission lines in or near indigenous lands (Couto, 1996).[65]

Among the main direct and indirect interferences arising from electricity sector expansion in indigenous areas, the following events stand out: relocation of communities to other regions, often accompanied by disruptions in their lifestyles; flooding large land parcels, including areas considered sacred, such as traditional burial sites; the invasion of indigenous lands, indirectly facilitated by the expansion of installations in the electricity sector; decreases in the availability of game and the reduction of arable areas; and proliferation of the vector populations of various diseases, including arthropods and mollusks, leading to an increase in the incidence of malaria and other infectious diseases (Couto, 1996).[66]

---

[63] Brasil (n 60).

[64] Ibid.

[65] Couto, R. C. S., 1996. Hidrelétricas e Saúde na Amazônia: Um Estudo sobre a Tendęncia da Malária na Área do lago da Hidrelétrica de Tucuruí, Pará Tese de Doutorado, Rio de Janeiro: Escola Nacional de Saúde Pública, Fundação Instituto Oswaldo Cruz.

[66] Ibid.

According to data provided by the National Indian Foundation (FUNAI) (personal communication), there were 156 indigenous areas where some type of dispute had arisen as a result of clashes with the electricity sector (Table 1). These had the following geographic distribution: 39 in the State of Amazonas; 22 in Mato Grosso; 21 in Pará; 14 in Mato Grosso do Sul; 13 in Rondônia; 8 in Rio Grande do Sul; 7 in Paraná; 6 in Tocantins; 4 each in Pernambuco, Bahia, and Santa Catarina; 3 in Minas Gerais; 2 each in Alagoas and Maranhão; and 1 in each of the states of Goiás, Espirito Santo, Paraíba, Roraima, Sergipe, São Paulo and Amapá. This shows not only a nationally distributed problem, but also a concentration of these type of issues in the Northern Region. Here, more than 65% of disputes referred to by indigenous communities were with the electricity sector.,[67,68]

These populations and communities are vulnerable to being directly or potentially affected by the expansion of this sector. In total, around 117,837 indigenous people, according to FUNAI statistics, representing approximately one third of the entire indigenous population in the country have been affected (MS, 2000).[69] The main factor involved concerns construction of hydroelectric plants, accounting for 104 (67%) of documented interferences with indigenous areas, 26 of which are in operation. The second most frequent cause of interference was the implementation of high voltage electrical transmission lines inside indigenous lands (26 reports), followed by five reports referring to their proximity to areas bordering indigenous communities (Koifman, 2001).[70]

10.3.2 Public Consultations in Accordance with Brazilian Law

In order to carry out energy service regulation, from 1996 onwards, federal regulatory agencies were created in Brazil. These special autarchies oversee and regulate the provision of energy, by the private sector,

---

[67] Koifman (n 62).

[68] Funai, 1996, personal communication.

[69] [Citation missing].

[70] Koifman (n 62).

regarding public or public interest services. These agencies were inspired by British and American regulatory institutions and are endowed with administrative autonomy with democratic participation of society as a functional requirement. However, not all characteristics foreseen for one agency could be expected or foreseen in others.[71]

Public participation and consultations in accordance with Brazilian energy sector legislation stands out for the legal channels which offer social control and societal participation in the formulation of agency rules and decisions.,[72,73] These channels are available on their websites and are effectively used in public consultations carried out via agencies (Wimmer & Peixoto, 2008).[74]

Other examples of public consultations are carried out by ANEEL. These agencies receive suggestions and comments for proposed actions in progress. Public participation is via consultation material available online. There is also the possibility of sending contributions directly through Internet.[75]

Furthermore, public hearings are instituted by ANEEL whenever a matter implies changes or adjustments in the agency's legislation, or directly interferes with the interests of society and agents in the electricity sector. Therefore, information about public hearings is part of its decision-making process and is considered by the agency itself as important factors in transparency of its actions. The material regarding public participation is widely made available by the agency in the form of downloadable files. Individual or group contributions can be sent by the website itself, and then made available for download.[76]

---

71 Pinho, José Antonio Gomes de; Sacramento, Ana Rita Silva. Accountability: já podemos traduzi-la para o português? Rev. Adm. Pública, Rio de Janeiro, v. 43, n. 6, p. 1343–1368, 2009.

72 Musarra (n 38).

73 Costa (n 10).

74 Reguladoras: como se conjugam Autonomia Decisória, Especialização Técnica e Participação Popular na ANATEL. In: XXXII Encontro da ANPAD – EnANPAD, 2008, Rio de Janeiro. Anais... Rio de Janeiro: ANPAD, set.2008. 16 p. CD Rom.

75 National Electric Energy Agency. Normative Resolution n°827 of 2018. Amends Normative Resolution No. 583, of October 22, 2013, and makes other provisions. Brasília, 2018.

76 Ibid.

There is a preponderance of participants taking part in the relevant forums regulated by the agency (producers) against consumers and public authorities. This fact is explained mainly by the nature of the sector in question. Electricity regulation is a sector that imposes high benefits concentrated to agents with their costs somewhat diffused for most consumers (Wilson, 1989).[77]

This encourages producers to be more organized and defend their interests which can in turn lead to them suffering more from the problems of collective action (Olson, 1999).[78] When regarding documentation sent as subject suggestions for analysis, the result is even more informative. Of the total contributions from all participants, 73.47% were from the regulators and only 8.12% from consumers. The effects of the distribution of costs and benefits discussed earlier are felt even more strongly. This is mainly due to the asymmetry of information and resources committed by agents in the construction of the documents sent.[79]

Disseminating the results of the contributions analysis carried out by the agency, is a moment of greatest relevance as it attests to how effective a certain action was on the final result of the matter in question. We emphasize the relevance that such a mechanism has within the scope of ANEEL, since all matters submitted to public hearings were modified at the suggestion of a group.[80]

In this matter, there is the so-called democratic deficit of Brazilian regulators. The laws and decrees for the creation and regulation of regulatory agencies do not have a legal provision for social control for said agencies. Although the Internet is a powerful tool in promoting democratic and social control, not all federal regulatory agency websites provide the necessary information for society to exercise its rights.[81]

---

77 [Citation missing].

78 [Citation missing].

79 Silva, Mariana Batista. Mecanismos de participação e atuação de grupos de interesse. no processo regulatório brasileiro: o caso da Agência Nacional de Energia Elétrica (ANEEL). Available at: https://www.scielo.br/j/rap/a/4Q84QS4qfxXbsf9DYm65Z7D/?format=pdf&lang=pt. Accessed Jul 2021.

80 Ibid.

81 Pinho and Sacramento (n 71).

Controlling the activity of the State is one of the pillars that sustains democracy.[82] Demajovic (2015) stated that society's participation is a requirement for legitimizing regulatory activity.[83] Jouravlev goes on to says that "consumer participation in the regulatory process is essential to achieve efficient sustainable and equitable regulation."[84]

From the redemocratization of Brazil, which had as a major milestone the promulgation of the Federal Constitution of 1988 (CF/1988), there was an opening for the participation of society in the management of public policies. Article 37, paragraph 3, states that the law will regulate the forms of user participation.[85]

In view of the constitutional provision, Article 29 of Law No. 8.987, of February 13, 1995 (Law No. 8.987/1995), deals with the provision of public services by means of concession and permission. This legislation, among the general duties of public administration related to the theme, also encourages the creation of associations of service users, in addition to receiving, investigating, and resolving user complaints. This vision was reinforced shortly after by the Federal Government, with the publication of the Master Plan for the Reform of the State Apparatus. This document guided the reforms undertaken during the government of Fernando Henrique Cardoso and states that, with the reforms, the Government would pass and also seek "direct social control and the participation of society" in the provision of public services (Brasil, 1995).[86]

Meanwhile, the main objective was to protect public interest, which is the objective of regulatory agencies, against the possible excesses of the regulated company, i.e., the maximization of profits. Pinho and Sacramento remind us that social control must be applied both in the strategic

---

[82] Ibid, p. 1350.

[83] Demajorovic, Jacques. Josantiago, Ana Lucia. A licença social para operar: um estudo de caso a partir de uma industria brasileira de mineração social license to operate: A case study from a brazilian mining industry. Available at: https://www.engema.org.br/XVIENGEMA/199.pdf. Accessed May 2021.

[84] Jouravlev, Andrei et al. Revisiting privatization, foreign investment, international arbitration, and water. Available at: https://repositorio.cepal.org/handle/11362/6328. Accessed May 2021. p. 22.

[85] Governo Do Estado. Federal Constitution of 1988 (CF/1988). Available at: http://www.planalto.gov.br/ccivil_03/constituicao/constituicao.htm. Accessed in 12 Jun 2021.

[86] [Citation missing].

sphere of public administration and in the production of public goods and services, including those that have been privatized.[87] However, the asymmetry of information between citizens and the State, combined with low level of organization of civil society, enables the installation of a scenario of "institutional weakness."[88]

Peci (2004) states that "regulatory reform" was an important institutional innovation in the country's political context. It influences both public administration and the relations between State and Society, since it incorporates a regulatory role and more public administration to the State.[89] In addition, it brings social control to the discussion as part of the democratic and regulatory process. Although the first regulatory agency within the regulatory model currently adopted in Brazil was only created in 1996, before the administrative reform, other institutions already existed with regulatory authority, such as the Central Bank of Brazil (BACEN), the Administrative Council for Economic Defense (CADE) and the Brazilian Securities Commission (CMV).[90]

This regulatory competence was also provided for in Law No. 8.987/1995 (BRASIL, 1995), which, in its Article 29, gives the State a series of tasks related to the concession and permission for the provision of public services. This includes the regulation of services, the application of regulatory and contractual penalties, intervention in service provision, the review and approval of tariff readjustments, help deliver good quality service, handle user complaints and encourage competitiveness.[91] In the Ministry of Administration and State Reform (MARE) and the elaboration of the Master Plan for State Reform, of September 21, 1995 (BRASIL, 1995),[92] we have the milestones of the second administrative reform of the Brazilian State initiated in the government of Fernando Henrique Cardoso.

---

[87] Pinho and Sacramento (n 71).

[88] Ibid.

[89] Peci, A. Vieira, M. M. F. Demarcacao cientifica uma reflexão critica. Curitiba: Enanped, 2004.

[90] ibid.

[91] Brasil (n 86).

[92] Ibid.

After this, and between 1996 and 2005, ten regulatory agencies were created in Brazil, as special autarchies. Among these, we can highlight: the National Electric Energy Agency (ANEEL), created by Law No. 9,427, of December 26, 1996, regulated by Decree No. 2,335, of October 6, 1997 and linked to the Ministry of Mines and Energy; the National Agency for Petroleum, Natural Gas and Biofuels (ANP), created by Law No. 9,487, of August 6, 1997, regulated by Decree No. 2,455, of January 14, 1998 and linked to the Ministry of Mines and Energy (Câmara, 1995).[93]

ANEEL provides consultations by agents, consumers, and society and can be considered as a public consultation. Both ANEEL and ANP provide for public hearings. These sessions, hearings, and public consultations allow society to give their opinions or directly propose changes in the norms and policies that directly affect it and are therefore important means of participation and social control.[94]

In most agencies, the decision-making process that affects the rights of agents, economic or consumer, may or must be preceded by a public hearing. Thus, interested parties can forward their requests on bills or administrative decisions of the agencies. Considering that the ombudsman is an important channel of social control, it is possible to affirm that these federal regulatory agencies have mechanisms and channels for control and the participation of society.[95]

Later, Bill No. 3.337/2004 (Brasil, 2004) appeared as a regulatory framework for regulatory agencies. It foresaw, in addition to the ombudsman's office, a process of public consultation with the publication online of its results, in addition to assuring participants the right to a reasoned response by the agencies to the questions submitted. This bill has the premise of strengthening and making the role of regulatory agencies more legitimate, thus reducing the democratic deficit, and states that the bill provides greater standardization regarding the positive elements

[93] [Citation missing].

[94] ANEEL (n 14).

[95] Peci (n 89).

present in agencies, despite not including important issues related to the control and participation of society.[96]

Regarding online public participation, it was possible to identify the channels of control and social participation available on federal regulatory agency websites. It was found that agency websites provided diversified information about means of access to different channels of participation, including via telephone, chats, and other means of online access, addresses for correspondence and addresses for face-to-face services.[97]

In addition to the means of access to information and as social participation channels, websites have the potential to provide citizens with tools allowing online participation in public consultations that are in progress, amounting to a great advance for citizen inclusivity. According to Moutinho dos Santos, ANEEL has access links to public board meetings, public hearings, and public consultations.[98] The ANP also provides access links to public consultations, public hearings, and board meetings. Information on bids, contracts, rendering of accounts, and public tenders was easily found on both websites.[99]

With improvements in the access to technology, public consultation via the Internet may become a democratic form of social participation, as it allows the participation of citizens from any Brazilian region on matters proposed by regulatory agencies whose headquarters are in the Federal District.[100]

It is important to highlight that the mechanisms of participation and social control on regulatory agency websites are accessible and visible. This includes access to public consultations, public hearings, and the ombudsman in addition to other items related to accountability, such as

---

[96] Santos, Alexandre. A legitimação democrática das agências reguladoras. Revista de direito público da economia, v.2, n.6, pp. 9–26, abr./june 2004.

[97] Demajorovic (n 83).

[98] Moutinho Dos Santos, E.. *The Brazil and gas sector:* outlook and opportunities. 1. ed. Londres: CWC Publishing Ltd. And Imedia Printed, 2001.ONU PORTUGAL. Avaliable at: http://www.onuportugal.pt/final-direito.pdf. Acesso em: 16 out. 2016.

[99] Ibid.

[100] Goldemberg, José; Lucon, Oswaldo; 2006. *Energia e meio ambiente no Brasil*. Availaable at: http://www.scielo.br/scielo.php?script=sci_arttext&pid=S0103-40142007000100003&lng=pt&nrm=iso. Accessed 20 May 2021.

information on contracts, bids, accountability, and public tenders. The agency websites visibly show links to all or at least some of the channels and mechanisms of control and social participation.[101]

However, on the ANP website, it is necessary to register in order to access or participate in public hearings or consultations that are in progress. It should be noted that most contributions came from companies, unions, and associations, with very little participations from citizens not acting on behalf of a group. This suggests a reticent response by the general populace to public consultations, despite the ease of access through the Internet. It also shows that organized groups have a great influence over the matters dealt with by these agencies.

10.4 Final Considerations

Throughout this article, an urgent need for a better regulatory framework for public participation in Brazil was seen. This includes the corresponding steps: identification of critical points, the agents involved, as well as protecting of areas and communities involved. The different aspects of SLO and communities were pointed out, and the need and acceptance of the people to any energy project is key in any part of the world, especially in Brazil. The concept of social justice was described as both consultation and the benefits for the community, which are important, as is addressing people's needs in order to facilitate the energy transition and the concerns that have generated interest in social and environmental public policies inserted into the national oil and gas sector.

It was highlighted that when it comes to socio-environmental protection, it is possible to glimpse not only the environmental impacts and morphological changes in the landscape, but also how the biophysical structure can be affected by the introduction of turbines, the alteration and even destruction of vegetation cover and the possible disappearance of characteristic landscape elements and many other impacts. In this scenario, the Brazilian Law No. 12,187, of December 29, 2009,

[101] Ibid.

discusses important concepts like mitigation, adaptation, and vulnerability. Regarding public participation and consultation in accordance with Brazilian energy sector legislation, the legal channels of social control and society's participation in formulating agency rules and decisions stands out. These channels are available on their websites and are effectively used in public consultations carried out via agencies.

In this matter, there is the so-called democratic deficit of Brazilian regulators and the laws and decrees for the creation and regulation of regulatory agencies. However, these do not have a legal provision for the social control of all agencies. Although the Internet is a powerful way to promote democratic and social control, not all federal regulatory agency websites provide the necessary information for society to exercise its rights. Also, almost all contributions come from companies, unions, and associations, with very little participation coming from citizens who are not acting on behalf of a group, which suggests the general populace is reticent in regards participation in public consultations.

Finally, based on the contribution made to research in this work, it will be possible to develop further relevant studies, with more in-depth analysis. There is scope for additional research that specifically glimpses a norm for the chosen theme. Furthermore, the need to incorporate other ways of involving the general populace could be demonstrated, providing a way to a more widely used, effective public participation.

Acknowledgements The authors would like to thank the support of SHELL Brazil and FAPESP, through the "Research Center for Gas Innovation (known by the acronym in English, Research Center for Gas Innovation- RCGI)" (Fapesp Proc. 2014 / 50279-4), hosted by University of São Paulo, for the resources that made this study feasible; and to the University of São Paulo, on behalf of the Faculty of Economics and the Energy and Environment Institute, for the opportunity of knowledge exchange that made the data collection and elaboration of information possible. The authors are also grateful for the important strategic support provided by ANP (National Agency of Petroleum, Natural Gas, and Biofuels in Brazil) through the regulation to promote Research and Development activities.

11

Social License to Operate: A Review of the Colombian Case

Luis Bustos, Ana Cecilia Zapata Sánchez, and Luis Fernando Bastidas Reyes

11.1 Introduction

In this first section, we will establish a basic framework associated with the idea of the Social License to Operate (SLO), later describing the use and relationship that this concept has had in the past years with regard to the Colombian experience, and its influence over the mining-energy

L. Bustos (✉)
Department of Mining and Energy Law, Externado University, Bogotá, Colombia
e-mail: luis.bustos@uexternado.edu.co

A. C. Zapata Sánchez
Land Restitution Unit, Government of Colombia, Bogotá, Colombia
e-mail: ana.zapata03@est.uexternado.edu.co

L. F. Bastidas Reyes
Social and Indigenous Affairs Consultant, Bogotá, Colombia

G. Wood and J. F. Neira-Castro, *From Fossil Fuels to Low Carbon Energy Transition*, Energy, Climate and the Environment,
https://doi.org/10.1007/978-3-031-00299-1_11

sector. Like many ideas, it has evolved and developed as an academic concept, and as a consequence, it is possible to find several definitions and uses of the term. However, it is pointed out that the original use of the SLO was born out of a search for solutions to the social barriers faced by the extractive industries, especially from the mining and oil industries.

Its use has expanded to this day, covering other sectors of the economy such as energy production and agribusiness. Regarding its use, the term was first referenced by Jim Cooney in a meeting on Communities and Mining promoted by the World Bank in 1997, using the expression to describe "*the local risk management*".[1] If we look further for a definition, we will find several approximations. However, these all hold the common view that the SLO in its simplest form is intended to bind a company with an obligation that encourages social acceptance of its activities. This can also be interpreted as an *"unwritten agreement"* between the stakeholders and the firm in which social support is required to enable the legal operations of the company.[2]

Gunningham offers another definition stating that the SLO consists of a way in which a company is guided towards meeting certain social expectations and in a parallel way avoiding activities that are unacceptable within society.[3] A community may decide not to grant a SLO when some aspects of social capital are not adequate and balanced, allowing the SLO acquisition process to place importance not only on projects but also on community network.[4]

[1] Meesters, M., Wostyn, P., Van Leeuwen, J., Behagel, J. H., & Turnhout, E. (2021). The Social License to Operate and the legitimacy of resource extraction. Current Opinion in Environmental Sustainability, 49, 7–11 P 7.

[2] Heffron, R., Downes, L., Ramirez Rodriguez, O. M., & McCauley, D. (2018). The emergence of the 'Social Licence to Operate' in the Extractive Industries? Resources Policy, [101272]. ISSN 0301–4207, https://doi.org/10.1016/j.resourpol.2018.09.012.

[3] Gunningham, N., Kagan, R. A., & Thornton, D. (2004). Social license and environmental protection: why businesses go beyond compliance. Law & Social Inquiry, 29(2), 307–341 P 307.

[4] Boutilier, R. G. (2021). From metaphor to political spin: Understanding criticisms of the social license. The Extractive Industries and Society, 8(2), 100743 P 3.

The SLO owes its origin to the concept of Corporate Social Responsibility (CSR), and most authors agree that this lack of a legal regulation is the principal reason the SLO is granted by communities.[5]

In other words, the SLO is an external element not connected to the legal structure as may well be in the case of permits and licenses, and definitely can't be provided by the state; on the contrary, the only grantor of this license is society and communities.[6,7] Under the foregoing, it seems that from a doctrine perspective an agreement exists to locate the SLO far away from legal constructions.

The SLO could be located within one of the CSR strategies and thus improve the relationship with its operating environment. However, another option would see its insertion at the centre of its operation of social factors. In either case, it would be a relationship characterised by company and community participation.

However, some authors question the real disposition of companies when they must contrast the SLO and the real consent of a community, remarking that it is not a simple participation in the decision-making process; in practical terms, they emphasise the refusal of some firms to withdraw their activities in the event of community opposition.[8,9]

[5] Bustos, L., & Rodríguez, M. (2016). Licencia Social Para Operar. Comentarios desde la Participación Ciudadana y la Responsabilidad Social Empresarial. At: Diaz, R; Lugo, O; Rodriguez, M. Regulación de Servicios Públicos y Energía. Porrúa. Pág. 197–220 P 197.

[6] Wilson, E. (2016). What is the social licence to operate? Local perceptions of oil and gas projects in Russia's Komi Republic and Sakhalin Island. The Extractive Industries and Society, 3(1), 73–81.

[7] Parsons, R., Lacey, J., & Moffat, K. (2014). Maintaining the legitimacy of a contested practice: How the minerals industry understands its 'social licence to operate'. Resources Policy, 41, 83–90.

[8] Morrison, John. (2014). The Social License. How to Keep Your Organization Legitimate? Palgrave Macmillan, ISBN: 9781137370716.

[9] Gunningham (n 3).

11.2 Colombia and the Social License to Operate

In the case of Colombia, we will find that the SLO concept has evolved. In the nineties, Colombia, like other Latin American countries, begin to adopt a series of neoliberal economic policy measures to guarantee development of its economy. Within this, we find amongst other actions: the opening of markets, the liberalisation of international trade, the elimination of barriers to foreign direct investment, privatisation and deregulation, the reduction of the state and the protection of private property. All ultimately was applied the formula described by Williamson in the Washington Consensus.[10]

The result of the aforementioned measures generated a bonanza in certain regional sectors of the economy.[11] An example of this could be seen in the progress made within the South American mining sector. It was also under these neoliberal structures that the Colombian mining code (Law 685, 2001) was conceived.[12]

Colombia is governed by a constitution created in 1991 within which the involvement of various environmental, social and participation aspects were established.[13] The materialisation of the constitution has not been an easy task. For this reason, amongst others, the 1991 Political Constitution has not been able to respond to social expectations.[14]

Under these circumstances, which lasted until the beginning of the new century, it was very difficult to introduce or integrate social CSR policies that went beyond simple altruism. It seems that mining and energy sector companies did not have much interest in the strategic

---

[10] Williamson, J. (1997). The Washington consensus revisited. Economic and social development into the XXI century IDB, 48–61 P 58.

[11] Bridge, G. (2004). Mapping the bonanza: geographies of mining investment in an era of neoliberal reform. The Professional Geographer, 56(3), 406–421 P 412.

[12] Law 685, 2001, Available online at: http://www.secretariasenado.gov.co/senado/basedoc/ley_0685_2001.html

[13] Amaya, Ó. D. (2002). La Constitución Ecológica de Colombia. U. Externado de Colombia. P 143–210.

[14] Political Constitution of Colombia (1991). The Republic of Colombia. (1991) Available online at: http://wsp.presidencia.gov.co/Normativa/Documents/Constitucion-Politica-Colombia.pdf

application of these types of policy. In the case of the SLO, we must realise that it was, at this point in time, a new concept and barely under construction.

In 2014, the Colombian Minister of Mines and Energy, Amylkar Acosta decided to use the SLO concept. He indicated that even though companies had an environmental license granted by the state and procedures were established and applied for prior consultations with communities, the projects would end up being unfeasible if they did not develop a SLO with the communities. However, Minister Acosta indicated in his pronouncements that it was a figure that came from CSR.[15]

From this point on, the use of this term in the usual language of the mining, oil and energy sectors became more common. References to the SLO began appearing in state public policy documents, an example of this is the National Plan for Mining Management, which in its text indicates: *"Social license (or dialogue) for medium and large mining projects. This instrument is the space to unfold the potential of Corporate Social Responsibility, RSE, beyond poverty reduction"*.[16]

Another aspect to consider when analysing the use and development of the SLO concept in Colombia is the prevalence of this notion amongst debates. In particular, the participation of territorial local authorities in decisions relating to oil and mining projects takes place in their jurisdictions. This debate, ongoing between the years 2012 and 2014, greatly affected the national mining and oil industry agenda for several years.

Although the debates came out in support on an issue of participation in mining projects, the complexity of issues from a constitutional law perspective, such as territorial land use in the territory, quickly led to being worked through judicial decisions, leading in turn to the creation of several court pronouncements, in particular from the Constitutional Court which adopted a jurisprudential line through its decisions.

---

[15] Revista Semana. (January 12, 2014). Proyectos sin licencia son inviables. Revista Semana. Available online at: http: https://www.semana.com/pais/articulo/licencia-social-proyectos-mineros/190454/

[16] UPME. Unidad de Planeación Minero Energética. (2014). Plan Nacional de Ordenamiento Minero - PNOM. Available online at: https://www1.upme.gov.co/simco/PlaneacionSector/Documents/cartilla%20Plan%20Nacional%20de%20Ordenamiento.pdfjsp?i=330 P 27–28.

The participation of territorial local authorities in energy sector policy is a debate that goes beyond the mining and energy sectors. In fact, it reaches all the way to the foundations of the Colombian territorial organisation proposed in the 1991 Constitution.[17] Some of the judgements issued by the Constitutional Court in the 2014–2020 period, created to deal with the aforementioned problems, are as follows: (Constitutional Court, Judicial decisions C 123 - 2014, C 273–2016, C 035–2016, C 389–2016, T 445–2016, SU095–2018, C 053–2019, T 342–2019 and SU411–2020).[18,19,20,21,22,23,24,25,26]

Given the importance of the aforementioned debate, its evolution and the reality that it resulted in the provisional suspension of oil and mining operations in the country, the SLO concept may have been caught in the middle of this debate. This could have led to the generation of inaccurate interpretations about its use and implementation by companies, government and communities.

In the following paragraphs, we will deliver some notes on the Colombian Constitutional Court decision SU 095 of 2018. This unified criteria delivered jurisprudential rules in an attempt to offer solutions to the growing debate on land use planning, and the relationship between the mining and oil industry and the territory and its authorities.

The ruling endeavoured to maintain an open dialogue between the authorities, communities and entities in charge of environmental concerns. Furthermore, it sought the development of the mining and oil sectors where no nation or local authority development model existed.

---

[17] Montoya Pardo, M. F. (2021). Licencia social para operar en el sector minero y novedades en materia de participación ciudadana. (Externado University) Curso Relacionamiento Territorial, Comunidades & Sostenibilidad en La Industria Minera y Petrolera 2021, Bogotá, Colombia P 401–402.

[18] Constitutional court. (March 5, 2014). Judicial decision C-123 [M.P: Rojas, A.].

[19] Constitutional court. (May 25, 2016). Judicial decision C-273 [M.P: Ortiz, G.].

[20] Constitutional court. (February 8, 2016). Judicial decision C-035 [M.P: Ortiz, G.].

[21] Constitutional court. (July 27, 2016). Judicial decision C-389 [M.P: Calle, M.].

[22] Constitutional court. (August 19, 2016). Judicial decision T-445 [M.P: Palacio, J.].

[23] Constitutional court. (October 11, 2018). Judicial decision SU-095 [M.P: Pardo, C.].

[24] Constitutional court. (February 19, 2019). Judicial decision C-053/19 [M.P: Pardo, C.].

[25] Constitutional court. (July 30, 2019). Judicial decision T-342 [M.P: Guerrero, L.].

[26] Constitutional court. (September 17, 2020). Judicial decision SU-411 [M.P: Rojas, A.].

However, at the same time, it proposed that these local authorities cannot have veto power over the activities. This pronouncement sustains the use of the principles of coordination and concurrence in these type of decisions.

Following the same logic, the aforementioned decision proclaims respect for the functioning of the organs of state and the decision-making process. It also prohibits any business activities from being excluded by a local decision, furthermore indicating that local authorities and communities must respect institutional channels giving. Furthermore, it gives relevance to the inbuilt democratic tools gained through the election of representatives, and thereby not allowing objectives to be achieved through unlawful conduct "*vías de hecho*".

Regarding the action of the national authorities, the pronouncement asks for the respect of local territorial authorities, avoiding decision-making without first listening to the territory, thereby strengthening the work of these entities.[27] The Constitutional Court also attributes great importance to the delivery of information transparency by all parties. This helps create a constructive dialogue, allowing clear communication and the fostering of trust, supporting citizen participation in decision-making. Additionally, it highlights the importance of protecting the rights of people involved in this type of participation process.

This ruling changed the way areas in the country were allocated for exploitation of their natural resources. Moreover, it has been reflected in public policies and actions from agencies in charge of the administration of the natural resources. Likewise, the Constitutional Court requests the executive and legislative branches to establish a legal norm where rules are unified and participation of communities is defined in this type of project. Currently, this rule is not part of the Colombian legal system.

As much as the described debate was decisive in developing the country's mining and energy industries, it cannot be constituted as the determinant of CSR policies linked to the SLO. In other words, the SLO

[27] Martinez, A. (2021). Instrumentos de ordenamiento territorial y competencias de las autoridades frente al desarrollo de actividades mineras y petroleras: Nuevos desarrollos. (Externado University) Curso Relacionamiento Territorial, Comunidades & Sostenibilidad en La Industria Minera y Petrolera 2021, Bogotá, Colombia

involves other issues that must be considered by companies operating in Colombia.

If twenty years later we returned to the contribution Jim Cooney gave us in his introduction to the concept of SLO, it would seem that the interpretation given by most Colombian stakeholders fails to consider an important point; central to the SLO idea is the issue of political risk of the undertaken projects and, that according to the above, two possible ways must be taken into account.[28]

The first of these deals with government permits required in order for operations to proceed. Some of these are issued for work related to risk reduction safety, environment and health[29] (Cooney, 2017 p. 198). The second is characterised by recognising the effects that globalisation has had on local communities, in particular how this has provided them with knowledge about industry projects and how, as a result, their bargaining power has increased. Within this view comes to light what could be described as a kind of support or sympathy that communities might find in other sectors of society, where they could make alliances and strengthen their negotiating position with companies.[30]

The union of these two different paths has the effect of strengthening the negotiating power of the communities. Problem arise when the SLO is used incorrectly and legitimises a form of activism that ignores judicial powers, the legislative structures and the government itself. This can lead to a loop where the disregard of the republican institutions becomes a determining factor in the increase of social demands through unlawful or improper conduct,[31] thus generating a greater or progressive imbalance in the democratic system.

At this point, we feel there is a need to evoke the words of Professor Jeremy Waldron who indicates: *"But the Rule of Law is not just about government. It also requires that citizens should respect and comply with*

---

[28] Canada Science and Technology Museum (January 09, 2017). Social License to Operate - Jim Cooney. [video file]. Available online at: https://www.youtube.com/watch?v=Mwbw9KF-ACY

[29] Cooney, J. (2017). Reflections on the 20th anniversary of the term 'social licence'. Journal of Energy & Natural Resources Law, 35(2), 197–200.

[30] Ibid P 198–199.

[31] Lucero, Luis E. (2021). Social Licence and the Rule of Law. Lexology Practice Guides, Mining. Available online at: https://www.lexology.com/library/detail.aspx?g=a4a96d4e-0d3d-44c4-bfd1-6abcf156a0ab

legal norms, even when they disagree with them".[32] In the same sense, the SLO may have been used to disregard the importance of the rule of law in the democratic system leading to an inappropriate interpretation of the theory.

The social aspects should be considered as a determining element within the policy and regulatory changes that are coming from the modifications in energy sources. Therefore, policymakers should not ignore environmental aspects; additionally, it is necessary to include social subjects and give them recognition in the construction of tools that support the process.

It is usual that environmental and social goals share similar commitments and complement each other; however, there are cases where tensions appear. Such an example is the conservation of moorlands where, on the one hand, the protection of the supply and production of water is sought, but on the other, there are economic activities that must be displaced from these protected areas on which the income and well-being of families depend.

In the following sections, we will propose two topics where the SLO would fulfil a fundamental role in the future construction of the Colombian mining and energy industries.

11.3 Colombia's Lands Restitution Public Policy and the Relationship with the Concept of Social License to Operate

The extractive industries in Colombia should not be considered too powerful to fall outside the kind of territorial and community issues that the country has faced for more than five decades. Indeed, there are multiple topics and aspects surrounding these industries that could be looked at. These include human rights, poverty, ethnic minorities,

[32] Waldron, J. (2016). The rule of law. The Stanford Encyclopedia of Philosophy (Summer 2020 Edition), Edward N. Zalta (ed.), Available online at: https://plato.stanford.edu/entries/rule-of-law/#pagetopright

environmental risks, infrastructure, lack of home public services, strikes, armed conflict, antipersonnel mines and the signing of the peace agreement in 2016, amongst others. This chapter looks to highlight key elements of the lands restitution public policy that somehow has come to influence the main purpose and spirit of social licenses to operate.

In general, to develop a project, a geographic location must be defined, bounded and explored. However, in the case of Colombia where social and political contexts are more complex (World Bank, 2021),[33] deciding exactly the area where each project is going to take place has become more difficult due to the consideration of several new non-technical variables. Proper due diligence in defining which land to use and which communities or interest groups to talk to is highly recommended during the planning stages and before going any further with the execution of projects. Decisions taken without factoring in the territories and communities could affect the viability, budgets and even the project schedule.

Since the year 2011, Colombia has for the first time carried out a public policy of land restitution, for the benefit of the victims of violence and armed conflict. According to, 1448 of 2011,[34] Law 2078 de 2021[35] and other complementary decrees, special procedures have been adopted that allow a person or ethnic group (indigenous, African descendants or gypsy) that have been victims of armed conflict can request a return to the lands they lost due to dispossession or forced abandonment. The procedure takes place in two main stages and in the following order: an administrative stage before the "*Unidad Administrativa Especial de Gestión de Restitución de Tierras Despojadas*"[36] and then a judicial stage before special judges and magistrates. One important requirement for

[33] World Bank. (2021). The World Bank in Colombia - Overview. Available online at: https://www.worldbank.org/en/country/colombia/overview

[34] Law 1448, 2011, Available online at: http://www.secretariasenado.gov.co/senado/basedoc/ley_1448_2011.html#:~:text=Por%20la%20cual%20se%20dictan,y%20se%20dictan%20otras%20disposiciones.

[35] Law 2078, 2021, Available online at: http://www.secretariasenado.gov.co/senado/basedoc/ley_2078_2021.html#:~:text=OBJETO.,2011%2C%20y%204635%20de%202011.

[36] Unidad de Restitución de Tierras (URT). Available online at: https://www.restituciondetierras.gov.co/

the success of these procedures is that the reasons that caused the dispossession or forced abandonment must have taken place after the 1st of January 1991.

It is important to highlight that the term dispossession is defined by the mentioned Law 1448 of 2011 article 74 as the action by which taking advantage through the situation of violence, a person is arbitrarily deprived of his or her property, possessions or occupation, whether, in fact, through legal business, administrative action, judgement or by committing crimes associated with the situation of violence.[37] Likewise, the term forced abandonment is defined by the same law in article 74 as "a temporary or permanent situation to which a person is forced to move, which is why they are prevented from exercising administration, operation, and direct contact with the land which they had to neglect when travelling during the period laid down under article 75".

Considering that the main purpose of this public policy is to give lands back and restore rights, case studies show that some trials involve only one party (the person or ethnic group). Other trials involved two or more parties (the person or ethnic group, various other applicants, the current person or ethnic group living in the land, the previous owner or owners, etc.) (URT, 2021b).[38] The diversity of scenarios that arose in the first decade of the implementation of this public policy (2011–2021) has been translated into a multiplicity of scenarios that extractive industries must overcome whenever they approach territories and communities.

In May 2021, the *"Unidad Administrativa Especial de Gestión de Restitución de Tierras Despojadas"* published new statistics that include a total of 101,492 people who had requested restitution of lands and over 116,674 sites or properties. Regarding the ethnic groups, 254 cases are currently in the administrative stage and 128 cases are currently in the judicial stage. Figures will continue increasing, since the public policy was extended for 10 more years, until the 10th of June 2031 (Law 2078, 2021).

[37] Law 1448 (n 24).

[38] Unidad de Restitución de Tierras (URT). (2021). Sentencias por Departamento. Available online at: https://www.restituciondetierras.gov.co/sentencias-por-departamento

In retrospect, the first ten years of the lands restitution public policy (2011–2021) has exposed many challenges and questions marks for both the public and private sectors. One of the most common questions is: What to do if a project requires the use of land that is totally or partially under a restitution process(es)? The answer will depend on multiple factors such as: the type and phase of the project (exploration, exploitation), the type of use required (permanent, temporary), duration (how long the land will be used), the type of community requesting the land, the stage of the restitution process and any other restrictions determined by law, etc. Other frequently asked questions are: Should a proper social license to operate procedure consider all parties of the restitution process to maintain its legitimacy? If the restitution process is not over, and there is no certainty who the owner of the land is, who should address the social license to operate procedure? Can new restitution cases that currently do not include all the parties affect previous social licenses to operate?

Nevertheless, we are facing both the social licenses to operate and the lands restitutions cases, as new chapters in the extractive industries handbook. Up to this point, the question was: Should we put the social licenses to operate chapter before the lands restitutions cases chapter or vice versa?

Given that the nature or source of lands restitution cases is precisely a breach in human rights protection and loss rights, social licenses to operate procedures would be wrong if they promoted a similar breach or loss by leaving out members of the communities or worst still, whole communities. Having said that, it seems more accurate to incorporate all the results of a proper lands restitution cases analysis into social licenses to operate schemes. In other words, the location of a project will allow us to identify the territories and their communities. This is going to lead to specialised topics such as lands restitution cases analysis the product of which could be the base to process social licenses to operate. The above could be seen as a fractional process, but it could happen so fast that may seem simultaneous.

Considering that any interaction with communities requires correct and responsible identification of its members, a key factor of social licenses to operate is to involve all parties, especially those who at first glance are unseeing and unheard. Therefore, because of all the new dynamics that the land restitution public policy has created over the country, it has become necessary to integrate this topic into the company decision-making process and in the variables of local and central authorities that control or administrate extractive industries projects.

Even though territories and communities are two different things, the national context implies that both concepts always go together. Proper identification of the land required for the projects will result in knowing the territory, its communities and its dynamics. All these are of utmost importance because they will contribute essential raw material to the design of a proper, solid social license to operate strategy.

Within the process of knowing the territory, proper identification of restitution cases is vital. Case logging should include, amongst other factors, the stage (administrative or judicial); type of community involved (farmers or ethnic groups); the number of communities; and environmental restriction.

11.4 Prior Consultation

Convention 169 of the International Labour Organisation (ILO 169) establishes the right to prior consultation of native and tribal people due to legislative or administrative measures that may affect them directly.[39] However, the Colombian legal system extends prior consultation to all types of measures, projects, works or activities likely to directly affect ethnic groups, that is, native, black, afro-descendant, *"raizales, palenqueros"* and gypsy groups, regardless of whether these are in charge of public or private entities.

This convention was incorporated into the Colombian legal system through Law 21 of 1991.[40] Even though it does not define prior consultation as a right or as a procedure, it includes the general conditions

[39] Convention No. 169 on native and tribal peoples in independent countries is approved, adopted by the 76th. Meeting of the ILO General Conference, Geneva 1989.

[40] Law 21, 1991. Available online at: http://www.suin-juriscol.gov.co/viewDocument.asp?ruta=Leyes/1577376

and purposes with which the consultations should be carried out when applying them.

From the convention, it can be concluded that prior consultation i) is a duty of the governments adhered; ii) the interested parties are the right holders, that is, those tribal or native people provided for in Article 1 of the text who have the will to be consulted; iii) it is performed through appropriate procedures; iv) it is performed through the representative institutions of the people concerned; v) it must be done in good faith; vi) it must be carried out properly to the circumstances; and vii) it has specific purpose.

The purposes of consultation provided for in ILO 169 establish the guidelines for the right to prior consultation and consequently the administrative and judicial decisions taken on that matter. ILO 169 paragraph 2 of Article 6 indicates that the prior consultation has the purpose of (i) reaching an agreement or (ii) obtaining consent about the proposed measures. Concerning prospecting and exploitation of the resources existing in the subsoil owned by the State, Article 15 of the ILO 169 includes a third purpose consisting of: (iii) determining whether and to what extent the interests of ethnic people would be harmed.

For the latter case, it is inferred that such purposes are fulfilled when: (i) consent is obtained regarding the proposed measures, or when, not having achieved it; (ii) a total agreement is reached on the proposed measures and it is determined whether the interests of ethnic groups would be harmed, and to what extent; (iii) a partial agreement is reached on the proposed measures and it is determined whether and to what extent the interests of ethnic groups would be harmed; (iv) a partial agreement is reached on the proposed measures and it is not determined whether and to what extent the interests of ethnic groups would be harmed; (v) no agreement is reached on the proposed measures, nor is it determined whether and to what extent the interests of ethnic groups would be harmed.

The right to prior consultation is not provided in the 1991 Political Constitution. However, its Preamble and Articles 1, 2, 7, 13, 40–2, 329, 330 and Transitory 55 are in harmony with the purposes of non-discrimination, integration and protection of native people and tribal communities and their cultural diversity provided by the ILO 169. The

ILO 169 is integrated into the constitutionality block and therefore has constitutional status.[41]

It is important to highlight that article 329 of the Colombian Constitution grants the right to the formation of native territorial entities whose delimitation is carried out with the participation of representatives of native communities. Article 330 grants the right to native territories to be governed by councils formed and regulated according to the uses and customs of their communities. Likewise, it indicates that the exploitation of natural resources in native territories is carried out without detriment to the cultural, social and economic integrity of native communities. Similarly, decisions that are adopted regarding said exploitation, the Government will encourage participation of representatives from the respective communities.

Colombia is a democratic, participatory and pluralist republic that defines as an essential purpose of the State to facilitate participation of all interested parties in decisions that affect them and in the economic, political, administrative and cultural life of the nation. Participation permeates the entire legal system, constituting a cornerstone of the political structure,[42] and is an instrumental pillar for the realisation of rights. The participation of ethnic groups is aimed at the preservation and defence of multiple rights, especially those related to their connection with the environment. Therefore, the impact of these is very important regarding the achievements in legislative and administrative decision-making or the implementation of projects, works or activities,[43,44] and for the nation's survival and ethnic and cultural diversity.

Even though there is no prior consultation law in Colombia, different legal provisions exist including the rules of the game on this matter, but lacking adequate consistency between them. Therefore, there is no absolute precision of the content and scope of the right of prior consultation

---

41 Political Constitution of Colombia (n 14).

42 Macedo Rizo, M. (2010). Citizen participation in the 1991 Constitution. Available online at: https://revistas.unal.edu.co/index.php/peju/article/view/36714/39502

43 Rodríguez, G. A. (2014). From prior consultation to free, prior, and informed consent to native people. Bogotá: GIZ, German Cooperation and Universidad del Rosario.

44 Rodríguez, G. A. (2015). Native people rights: Struggles, content and relationships. Bogotá: Editorial Universidad del Rosario.

and its application mechanism. It is worth clarifying that there is no legal or regulatory definition of prior consultation.

The domestic legal system is made up of different legal sources regarding prior consultation, namely: (i) Law 21 of 1991 approving ILO convention 169, by which it is incorporated into the Colombian legal system[45]; (ii) different legal norms that invoke, mention or order the right to prior consultation but do not regulate it, as well as its implementation procedure; (iii) Decree 1320 of 1998,[46] which even though it has not been annulled by the competent judicial authority, was declared inapplicable by the Constitutional Court by means of Sentence T-652 of 1998.[47] Therefore, its application at present is strange and problematic, and therefore, it is not applied by the administrative authorities; (iv) Decree 2613 of 2013[48] by which the inter-institutional coordination protocol for prior consultation is adopted, which, even though it is in force, its application is problematic since it is not compatible with the current competences of the National Prior Consultation Authority Direction (NPCAD) indicated in Decree 2353 of 2019[49]; (v) three presidential directives (01 of 2010, 10 of 2013 and 08 of 2020) that the constitutional jurisprudence and the Colombian Council of State recognise their binding power and legal effects on public entities, ethnic groups and the private sector; (vi) international instruments complying with the content of the right to prior consultation; (vii) Jurisprudence of the Constitutional Court and extension of jurisprudence of the Council of State.[50,51,52]

---

45 Law 21 of 1991 (n 40).

46 Decree 1320, 1998. Available online at: http://www.mininterior.gov.co/sites/default/files/co-decreto-1320-98-consulta-previa-indigenas-_2.pdf

47 Constitutional court. (November 10, 1998). Judicial decision T-652 [M.P: Gaviria, C.].

48 Decree 2613, 2013. Available online at: https://www.mininterior.gov.co/la-institucion/normatividad/decreto-2613-de-2013

49 Decree 2353, 2019. Available online at: http://www.secretariasenado.gov.co/senado/basedoc/decreto_2353_2019.html

50 Presidential Directive 01. (March 26, 2010). By which the general framework for compliance with ILO Convention 169 is established and the measures that may or may not be carried out by Prior Consultation are established.

51 Presidential Directive 08. (September 9, 2020). Guide for conducting Prior Consultation.

52 Presidential Directive 10. (2013). Step 2: Development of meetings to formulate agreements.

As legal provisions that regulate prior consultation, the Presidential Directives 10 of 2013 and 08 of 2020 stand out.[53,54] These divide the prior consultation process into 5 stages and establish rules of the game for each one: (1) determination of the origin of prior consultation; (2) coordination and preparation; (3) pre-consultation; (4) prior consultation; and (5) follow-up of agreements. Depending on whether the prior consultation is carried out regarding administrative and legislative measures or projects, works or activities, these stages may have some differences. Below we will refer only to prior consultation of projects, works or activities.

In the *first stage*, the executor of the project, work or activity requests the (NPCAD) to determine the origin of the prior consultation. For this, the possibility is considered that ethnic groups are susceptible to direct effects that compromise their survival and ethnic and cultural diversity. If the prior consultation is appropriate, it will issue an administrative act that will impose the obligation to carry out the respective prior consultation process and will indicate the ethnic groups that are the object of the same. Only the executor or person responsible for the works programme (called the POA) can be an applicant for the determination of origin or not prior to consultation with the NPCAD, as the analysis of possible direct impacts takes place in consideration of the activities description of the mentioned programme, and for which they are eventually responsible to other administrative authorities.

In the *second stage*, it is the responsibility of the NPCAD, together with the executor of the project, work or activity, to identify the public entities whose competencies are linked to its execution and who to be included in follow-up meetings and designs strategies that facilitate the consultation process, and if possible, identifies whether obtaining prior, free and informed consent is required.

In the *third stage*, the NPCAD and executor of the project, work or activity carry out a dialogue with the ethnic groups under consultation through their representative authorities, to obtain consent. If this is not

53 Ibid.

54 Presidential Directive 08 (n 51).

achieved, the methodological route that they will jointly follow during the following stages is defined.

In the *fourth stage*, the executor of the project, work or activity of the project and the ethnic groups identify the possible impacts and formulate management measures. Finally, the agreements are formulated and formalised in writing.

In the *fifth stage*, the executor of the project, work or activity and the ethnic groups verify that those agreements formalised in the fourth stage are fulfilled.

The foregoing allows the prior consultation to be conceived as a procedural legal institution.

This constitutes a sensitive and current legal institution in the national debate. The resistance of ethnic groups and the impetus to conserve territories under their occupation in Colombia against the implementation of projects, works or activities involving the exploitation of natural resources, especially those of an extractive nature, have led the judiciary to adopt strong intervention measures in the origin of the prior consultation in Colombia.

The Constitutional Court, based on the study of particular cases, has defined a series of rules on the matter, guiding the actions of the parties involved in the prior consultation process. Two main points stand out:

(a) Prior consultation proceeds whenever there is the possibility of direct concern for the ethnic group. Under Decree 2353 of 2019, the NPCAD is responsible for its determination. If its decision is affirmative through an administrative act, it will indicate the ethnic groups' communities of which the prior consultation must be carried out. In the case of projects, works or activities, it is up to the executor to specify the circumstances of time, manner and place in which they will be developed. Regarding these, the NPCAD will evaluate the susceptibility of causing direct effects on ethnic groups. It is important to indicate that in judgement SU-123[55] certain exceptional cases were mentioned in which an *"intense direct impact"* was identified

[55] Constitutional court. (November 15, 2018). Judicial decision SU-123 [M.P: Rojas, A. & Uprimny, R.].

that requires obtaining prior, free and informed consent, since "... *if the agreement is not reached, the protection of traditional communities will prevail...*". These are cases that are based on jurisprudential developments and international law and will be applied when faced with: i) a transfer or relocation of native or tribal people from their place of settlement; ii) the storage or deposit of dangerous or toxic materials in their territories; iii) certain measures that imply a high social, cultural and environmental impact that puts their subsistence at risk.[56,57,58,59,60,61,62,63,64,65,66,67]

(b) Prior Consultation does not admit any veto by the ethnic groups, but neither an arbitrary imposition of power by the State (Constitutional Court, Judicial decision SU-123, 2018), "(iv) in this intercultural dialogue, neither the people have a veto right nor the State an arbitrary power to impose the planned measure".[68]

This assumes that it is not a mechanism by which ethnic groups decide or authorise the implementation or not of the measure or project, work or activity. This corresponds to the administrative authorities subject to the law. Thus, it does not constitute a negotiation mechanism either, since it does not operate as an exchange of rights between the participants in exchange for a price. It is the law and its regulations that grant the right to public and private to carry out activities in Colombian territory.

The good faith and the non-existence of the right to veto over the different works programmes are guidelines for the exercise of the prior

[56] Constitutional court. (March 3, 2011). Judicial decision T-129 [M.P: Palacio, J.].
[57] Constitutional court. (October 29, 2009). Judicial decision T-769 [M.P: Pinilla, N.].
[58] Constitutional court. (May 18, 2012). Judicial decision T-376 [M.P: Calle M.].
[59] Constitutional court. (February 13, 2013). Judicial decision C-068 [M.P: Guerrero, L.].
[60] Constitutional court. (December 16, 2015). Judicial decision T-766 [M.P: Mendoza, G.].
[61] Constitutional court. (June 11, 2014). Judicial decision C-371 [M.P: Pretelt, J.].
[62] Constitutional court. (December 16, 2014). Judicial decision T-969 [M.P: Ortiz, G.].
[63] Constitutional court. (April 26, 2016). Judicial decision T-197 [M.P: Palacio, J.].
[64] Constitutional court. (December 16, 2015). Judicial decision T-764 [M.P: Mendoza, G.].
[65] Constitutional court. (October 23, 2015). Judicial decision T-661 [M.P: Calle M.].
[66] Constitutional court. (August 26, 2015). Judicial decision T-550 [M.P: Ávila, M.].
[67] Constitutional court. (May 5, 2015). Judicial decision T-256 [M.P: Sáchica, M].
[68] Constitutional court (n 55).

consultation right. But, they are definitely insufficient when it is necessary to attend to numerous circumstances that require legal certainty.[69] Likewise, in the guide on the application of ILO 169, prepared by the International Labour Organization, there is not a mention in which the prior consultation grants a veto power in favour of communities on the development policies of each country.[70] We mention this without ignoring that at the international level and mainly in sources of soft law, ambiguities have been incurred regarding the veto power of communities. Therefore, we must highlight it as a discussed aspect without affecting the situation of the Colombian internal order.[71]

Regardless, we consider that all game rules must be linked to the aforementioned purpose of prior consultation. It is insisted that they guide and provide a framework for action aimed at exercising the right to prior consultation. Consequently, the work of the Constitutional Court is transcendental, since game rules that go beyond or do not correspond to the purpose of prior consultation provided for in the Agreement. Furthermore, they may distort its nature and negatively compromise the adoption of administrative and legislative measures and the implementation of projects, works or activities.

In specific cases, the Constitutional Court has subordinated prior consultation failures considering the ethnic groups' worldviews. However, these antecedents have been mistakenly used to keep open the prior consultation processes as long as the ethnic groups wish it to do so. Doubtlessly, this supposes an abuse in the exercise of the right that therefore undermines the other parts of the process.

Achieving the goals does not represent an imperative, since it is a conduct of means and not a result. They are developed to the greatest

---

[69] Bastidas, L, & Bustos, L. (2021). Derecho de Consulta Previa en Colombia, Comentarios Sobre el Sector Minero. At: Caldeira, L., & de Mattos Silva, T. Minería en tierras indígenas en Latinoamérica: Desarrollo y Medio Ambiente. EKLA Konrad-Adenauer-Stiftung. Pág. 105–132.

[70] International Labor Organization, (2009). Los derechos de los pueblos indígenas y tribales en la práctica: una guía sobre el convenio núm. 169 de la OIT. OIT, ISBN: 978-92-2-322379-3 Available online at: https://www.ilo.org/global/publications/books/WCMS_126163/lang--es/index.htm

[71] For more information on what is proposed here, see: Rodríguez Garavito, C., et al. (2010). Prior consultation with native people: the standards of international law. Colombia: Universidad de los Andes. P 80.

extent possible, but are exhausted without requiring them to obtain an exclusive result. Failure to achieve them does not imply the cessation of constitutional protection for natives and tribal people.

The Constitutional Court (Judicial decision SU-123, 2018)[72] stated that:

> If the consultation had been carried out in due form and there is no agreement, the State can implement the measure, through a reasoned decision, as long as its decision: (i) is devoid of arbitrariness and authoritarianism, (ii) is based on criteria of reasonableness, proportionality and objectivity regarding the duty of recognition and protection of the ethnic and cultural diversity of the Nation; (iii) take into consideration, as far as possible, the positions expressed by the parties, and especially the ethnic people, during the consultation; (iv) respect the substantive rights recognized in ILO Convention 169; and (v) provide effective, suitable and efficient mechanisms to mitigate the negative effects of the measure.

Finally, it is worth mentioning that for Colombia the NPCAD as part of the Ministry of Interior stands as the highest administrative authority in the prior consultation process. Currently, the NPCAD has more than 10,200 active prior consultation processes. It serves as a balance to intercultural dialogue and its tensions.

Although it can propose agreements between parties, it does not intervene in their decisions since it has no competence to impose them, much less it is responsible for their compliance. Therefore, they do not have the responsibility to resolve discussions of rights or principles regarding the implementation of projects, works or activities. Therefore, it must ensure that the measures must be proportionate, appropriate and necessary. It can apply instruments for weighing rights or principles when addressing these conflicts, such as the proportionality test and respecting the decisions that are within the scope of its competencies.

[72] Constitutional court. (November 15, 2018). Judicial decision SU-123 [M.P: Rojas, A. & Uprimny, R.].

11.5 Conclusions

- The interpretation and use of the SLO theory by the Colombian authorities and communities may have led to inappropriate use of the term, and it is important to frame the use of the theory within the boundaries of the CSR; otherwise, this will lead to the creation of false expectations and frustrations within the communities and society.
- Two public policies like the extractive industries and lands restitution are learning to work together, facing challenges, identifying common themes and always looking to make lasting contributions in our societies. Still, the challenges persist because of the current lack of articulation between authorities and policymakers.
- Social factors and more specifically the SLO must be taken into account by the regulator when setting a roadmap towards the transformation of the energy sector; otherwise, you will run the risk of receiving social opposition to any type of project even though these represent a more environmentally friendly option.
- The regulator must consider the social factors when setting a roadmap towards the transformation of the energy sector; otherwise, it will assume the risk of receiving social opposition to any type of project, even if it represents an option that protects the environment.
- Prior consultation is an opportunity for ethnic groups to have an impact on the projects, works or activities that are intended to be carried out in territories under their occupation when these may cause direct effects that do not admit impositions and arbitrariness of any of the involved parties. The proper use of this figure contributes to the construction of the SLO in the long term for any type of Project.
- The existence, presence, recognition and inclusion of people and their lands are basic attributes of the lands restitution public policy, as well as key factors for a transparent and prosperous SLO approach that will lead to the growth and improvement of the Colombian mining and energy industries operation.

12

Peruvian Regulations for Renewable Energy Resources: An Analysis of the Regime Based on Arbitration Awards

Carolina Meneses

12.1 Introduction

According to the National Energy Policy of Peru 2010–2040, an energy system that satisfies a country's demand for energy reliably, regularly, continuously and efficiently promotes sustainable development. This system is based on planning and continuous technological research and innovation.[1] According to this plan, Peru's objectives, among others, are to have a diversified energy matrix, with emphasis on renewable energy

[1] Republica del Peru, Decreto Supremo No. 064–2010-EM 2010.

C. Meneses (✉)
Independent Consultant, Universidad Peruana de Ciencias Aplicadas (UPC), Lima, Peru
e-mail: carolinamenesessotelo@gmail.com

G. Wood and J. F. Neira-Castro, *From Fossil Fuels to Low Carbon Energy Transition*, Energy, Climate and the Environment,
https://doi.org/10.1007/978-3-031-00299-1_12

resources (RER) and energy efficiency and develop an energy system with minimum environmental impacts.[2]

In the last 30 years, the country's electrical coverage has increased from 57% in 1993 to 91% in 2013, with social inclusion policies that allow expanding the percentage up to almost 100% in the following years.[3]

To promote energy efficiency, Peru's objective would be to have a diversified and competitive energy matrix by 2040, emphasising RER of at least 20%. In this context, since 2006, the Peruvian State has been providing an essential boost to RER sources. Non-conventional RER uses the inexhaustible flow of natural energy sources (sun, wind, water, plant growth, sea movement and others) to supply the growing energy demand.[4]

In 2006, Peru started implementing regulations to promote electricity generation with RER; since then, auctions have been the most important mechanism to grant electricity supply contracts.[5] The enforced regulation has helped increase electricity generation with RER from 0.002% in 2008 to approximate 5% in 2016. As of the implementation of Law 28,832 and Legislative Decree 1002, a total of 4 auctions have been conducted, and the last auction (in 2016) granted a total of 24 RER generation projects.[6]

Despite the apparent success of the auction regulation for the generation of electricity with RER, the Peruvian State has been defendant on several arbitration cases related to the RER regulation and auctions. Considering the importance of the RER regulation in the country and its significance to the energy transition facing climate change, this chapter

---

[2] *Ibid* Objectives 1 and 7.

[3] Ministerio de Energía y Minas (MINEM), 'Resumen Ejecutivo. Plan Energético Nacional 2014–2025' (2014). p. 8, https://www.minem.gob.pe/minem/archivos/2ResEje-2014-2025Vf.pdf

[4] Organismo Supervisor de la Inversión en Energía y Minería—OSINERGMIN, La Industria de la Energía Renovable en el Perú: 10 años de contribuciones a la mitigación del cambio climático. 2017. p. 16.

[5] Riquel Ernes Mitma Ramirez, 'Análisis de La Regulación de Energías Renovables En El Perú' [2008] Revista Derecho & Sociedad. p. 168.

[6] Organismo Supervisor de la Inversión en Energía y Minería—OSINERGMIN La Industria de la Energía Renovable en el Perú: 10 años de contribuciones a la mitigación del cambio climático. (n 4) p. 23.

aims to analyse the RER regulation, its incentives and promotion of electricity generation with RER, as well as some relevant arbitration cases where Peru acted as a defendant, to determine if some amendments to specific provisions of the contracts granted by Peru as result of the auctions are needed.

12.2 RER Regime

12.2.1 What is the RER Regime?

Since 2006, with Law No. 28832, the Efficient Generation Law ("Efficient Generation Law"),[7] auctions have become the primary mechanism for awarding electricity supply contracts in Peru.[8] This law considers contracts derived from bidding processes or auctions as one of the forms of sale of electricity destined for the public electricity service. Thus, the primary purpose of the bidding processes is the timely and efficient supply of electricity to the regulated market through long-term electricity supply contracts.[9]

According to the Efficient Generation Law, supply contracts executed as a result of a bidding process must contain the same terms and conditions as the winning proposals and are subject to certain conditions[10]:

a) Supply terms of up to 20 years and firm prices. None of which can be modified by agreement of the parties, unless authorised by the competent body, the Supervisory Agency for Investment in Energy and Mining (in Spanish *Organismo Supervisor de la Inversion en Energia y Mineria*—"Osinergmin"); and
b) Firm price update formulas.

---

[7] Republica del Peru, Ley No. 28832: Ley para asegurar el desarrollo eficiente de la generación eléctrica 2006.

[8] Mitma Ramirez (n 5) p. 168.

[9] Republica del Peru Ley No. 28832: Ley para asegurar el desarrollo eficiente de la generación eléctrica (n 7) Articles 3.2.b and 4.

[10] *Ibid* Article 8.

In 2008, the Legislative Decree for the Promotion of Private Investment for the Generation of Electricity with the Use of Renewable Energies, Legislative Decree No. 1002 ("RER Law"),[11] was enacted. RER Law's purpose is to guarantee the timely and efficient supply of electricity with incentives to promote private investment in electricity generation using RER, improve the population's quality of life, and protect the environment.[12]

The RER Law declared the development of electricity through the use of RER of national interest and a public necessity, determined the competent authorities, the promotion mechanisms and implemented, by the Efficient Generation Law, an auction system for the granting of long-term electricity supply contracts.

Regarding the auction process, the RER Law establishes that the participating bidders design the specific projects to offer the amount of energy required and its price. The best (successful) bidder is awarded and signs an electricity supply contract with the Ministry of Energy and Mines (MINEM). MINEM guarantees the purchase of all electricity generated for up to 20 years at the price proposed by the awarded bidder. In return, the awardee is assigned the risks of design, obtaining authorisations, permits and licences, construction, operation and maintenance of the project.

Also, in 2008, through Supreme Decree No. 050–2008-EM,[13] the Regulation of Electricity Generation with Renewable Energies was approved. This regulation details the general conditions applicable to the auction processes and the recovery of investment by the awarded bidders.[14] However, after the first auction of electricity supply with RER, the need to have regulation that clarifies the application of the mechanisms provided in the RER Law was noted.[15]

---

[11] Republica del Peru, 'Decreto Legislativo No. 1002, Decreto Legislativo de Promoción de La Inversión Para La Generación de Electricidad Con El Uso de Energías Renovables' 7, https://www2.osinerg.gob.pe/EnergiasRenovables/contenido/Normas/DL_No_1002.pdf%5Cn: https://www.minem.gob.pe/archivos/legislacion-9ozj22z9ap5zz33z-DL_de_promocion_de_la_inversion_para_la_generacion_de_electricidad_con_el_uso_de_energias_renovables_1002.pdf

[12] Republica del Peru, Exposición de motivos. Decreto Legislativo No. 1002 2008.

[13] Republica del Peru, Decreto Supremo No. 050–2008-EM 2008.

[14] Republica del Peru, Exposición de motivos. Decreto Supremo No. 050–2008-EM 2008.

[15] Republica del Peru, Exposición de motivos. Decreto Supremo No. 012–2011-EM 2011.

Consequently, in March 2011, Supreme Decree No. 012–2011-EM ("RER Regulation") was enacted, repealing Supreme Decree No. 050–2008-EM (together with RER Law, these constitute the "RER Regime"). Subsequently, in 2013, the RER regulation was modified to facilitate the execution of projects with RER.[16]

The general criteria in the design of the regulations applicable to the promotion of generation with RER in Peru consider, among others, the following aspects:

a) Large scale;
b) Efficient and effective auction[17]: That encourages competition between bidders, giving priority to those who make the best use (offer lower prices) of RER[18] sources and have the availability to execute the projects[19]; and
c) Maximising the benefits of energy consumers.[20] Given that the electricity consumers assume the cost overruns of the energy produced with RER via rates' surcharge, the maximum prices paid by the State must be respected, and these prices are defined objectively and with technical criteria.

12.2.2 How Does the RER Regime Work?

The RER regulation establishes the processes and procedures to call auctions, their development, and the award of the bid, among others.[21] The purpose of this regulation is that the auction process is as efficient

16 Republica del Peru, Exposición de motivos del Proyecto de Decreto Supremo que aprueba la modificación al Reglamento de la Ley de Promoción de la Inversión para la Generación de Electricidad con el uso de Energías Renovales y el Reglamento a la Ley de Concesiones Eléctricas. 2013.

17 Riquel Ernes Mitma Ramirez, 'Regulación de Las Energías Renovables En El Perú' (2013) 28.

18 Mitma Ramirez (n 5) 170.

19 Ricardo De la Piedra Calle, 'Una Mirada a La Regulación y Resultados de Los Proyectos de Recursos Energéticos Renovables En El Perú' Revista Advocatus 222, http://revistas.ulima.edu.pe/index.php/Advocatus/article/view/4481/4392.

20 Mitma Ramirez (n 5) 170.

21 Republica del Peru, Decreto Supremo No. 012–2011-EM 2011.

and transparent as possible and allows the selection of the best bidders in each process.

The most critical aspects of the auction process are as follows:

a) Determination and composition of required energy. The energy required is determined by MINEM and considers the national electricity consumption and the participation of RER generation.[22] MINEM defines the percentage of participation of each RER technology in the energy required.[23]
b) Information and registration system. Osinergmin publishes the documents of all the stages of each auction process.[24]
c) Bases. The bases of each auction process are prepared and approved by MINEM and are mandatory for all bidders. The bases establish: (i) the end date of the supply contract to be signed with the auction winner; and (ii) the types of applicable guarantees (seriousness of the offer during the auction process, and of faithful compliance with the execution of works following the award to the best offer).[25]
d) Call for auction. Osinergmin publishes the call for auction with the following information[26]:

    - The energy required by each type of RER technology;
    - Reference date of Commercial Operation Commissioning (POC) (in Spanish *Puesta en Operación Comercial*);
    - End date of the supply contract to be signed with the successful bidder; and
    - Relevant milestones in the auction process schedule.

e) Requirements to be a Bidder. Those interested in participating in an auction process must meet, among others, the following requirements[27]:

---

22 *Ibid* Article 4.

23 *Ibid* Article 5.

24 *Ibid* Article 7.

25 *Ibid* Article 8.

26 *Ibid* Article 9.

27 *Ibid* Article 10.

- Have acquired the bases;
- Provide date regarding the project to be auctioned (nominal power to be installed, plant factor, project's location and connecting rod, etc.),
- Provide the guarantee of the seriousness of the offer, and at the closing date (if awarded with the auction), the guarantee of faithful compliance; and
- Be bound to comply with the supply contract of RER by filing and obtaining a definitive RER generation concession.

f) Presentation of offers. The price presented by the bidders as an offer must include: the necessary investment costs for its connection to the National Interconnected Electric System (SEIN) (in Spanish *Sistema Eléctrico Interconectado Nacional*), the power committed to be installed, the annual energy offered, and the expected date of POC[28] (the date when the awarded bidder shall supply electricity generated with RER to the State).
g) Determination of the base rate. This is determined by Osinergmin and considers efficient investment, operation, and maintenance costs.[29]
h) Evaluation of offers and award of bids. This is carried out in strict order of the offers that do not exceed the base rate until the participation established in the bases is completed, according to the total energy required for each type of RER.[30]
i) Award certificate and closing date. After awarding the bid of an auction, an award certificate is signed, and such certificate contains (for each awarded bid) the following[31]:

   - Termination date of the supply contract;
   - Energy awarded;
   - Expected date of POC; and

[28] *Ibid* Article 12.
[29] *Ibid* Article 13.
[30] *Ibid* Article 14.
[31] *Ibid* Article 15.

- Awarded tariff (with the updated formula established in the auction bases).

On the closing date, compliance with the requirements for executing the supply contract of renewable energy between MINEM and the successful bidder resulting from the auction is verified.[32]

The supply contract executed after awarding the bid ("RER Supply Contract") sets out each party's obligations in accordance with the RER Regime. Thus, such RER Supply Contracts contemplate the contract's termination date, the amount of energy awarded, the scheduled date for POC and the awarded tariff. Likewise, MINEMs obligations, such as the purchase of all energy to be supplied, the tariff to be paid (awarded tariff), and the obligation to prioritise the dispatch and connection to the SEIN, are established. The awarded bidder contractually assumes its bid offer and undertakes, among others, to process and obtain the necessary titles and permits for the project's development, including the definitive generation or authorisation concession, as necessary, depending on the RER to be used.

RER Supply Contracts are administrative contracts regulated by their provisions and, in the event of legal voids, by the RER Law and Regulations, and, additionally, by the Peruvian Civil Code, approved by Legislative Decree No. 295 ("Civil Code").[33]

Furthermore, certain principles of the private investment promotion regime (legality, good faith, reasonableness, continuity of execution, flexibility, maintenance of economic-financial balance) and arbitration as a means of solving controversies apply to these contracts. Likewise, respect for the fundamental rights of freedom of business, private initiative, and freedom of contracting and property, among others, is guaranteed, always within the framework established by the RER Supply Contract, the consolidated bases of the auction, and the RER Regime.[34]

---

32 *Ibid*.

33 Republica del Peru, Decreto Legislativo No. 295, Código Civil 1984 Article IX, Preliminary Title.

34 Ramón Huapaya Tapia, 'Una Propuesta de Formulación de Principios Jurídicos de La Fase de Ejecución de Los Contratos Públicos de Concesión de Servicios Públicos y Obras Públicas de Infraestructura' (2013) 23 Ius Et Veritas 284, 328–329, http://revistas.pucp.edu.pe/index.php/iusetveritas/article/view/11973/12541.

In addition to the aforementioned regulatory framework, there are additional aspects related to general issues, such as the operation of the market, necessary titles and permits to be filed and obtained, and the obligations according to the type of activity carried out, which are subject to an applicable general regulatory framework to all electricity generators without exception, and detailed in the Electricity Concessions Law (RLCE) (in Spanish *Ley de Concesiones Eléctricas*) and its regulations.[35]

As indicated above, by executing the RER Supply Contract, the concessionaire companies (awarded bidders) assume the obligation to manage and comply with all the requirements to obtain the main qualifying title: the definitive concession, which allows the development of electricity generation with RER.[36]

Based on the preceding, it is possible to conclude that, to develop the activities framed in the RER Regime, concessionaire companies (awarded bidders) execute a RER Supply Contract and, as a consequence, assume the obligation to file and obtain the definitive concessions according to LCE and the RLCE.

Thus, with the execution of the RER Supply Contract, a completely different stage and relationship from the auction procedure begins. The relationship between bidders and an administration in charge of governing an administrative procedure with pre-established rules, such as those of the auction, disappears. The relationship becomes bilateral as of the execution of the administrative contract. Thus, the administrative contract establishes that obligations and risks are assigned/allocated to each party, depending on the public interest that the RER Supply Contract pursues.

[35] Mitma Ramirez (n 5) 169.

[36] Depending on the project's specifications, the main qualifying title can also be an authorization. However, for practical purposes we shall refer to this qualifying title as a definitive concession.

12.3 Critical Aspects of the RER Regime

12.3.1 National Interest and Public Necessity

With the publication of the RER Law, considered a milestone in the generation of clean energy in the region,[37] there is a more significant boost to the promotion of RER. To promote investment in this type of energy, the development of electricity generation through RER was declared of national interest and public necessity.

Administrative or regulated contracts are governed by the public interest, which takes precedence over private interest and any other extra-legal consideration that has been taken into account at the time of entering into the contract. Effectively, this means that the parties in this legal relationship are essentially unequal, since the call to guarantee and monitor the protection of the public interest of the country and its population is the State.[38]

As a consequence of the preceding, within the framework of administrative contracts, the so-called exorbitant clauses are usually granted in favour of the public administration. The State can unilaterally modify the contracts it has entered into with individuals.[39]

In addition, according to the Peruvian Constitutional Court, the satisfaction of the public interest constitutes one of the aims of the State.[40] Thus, it corresponds to the State to satisfy such interest, for which it must ensure that the obligations of the bidders of the auctions established in the RER Supply Contracts are fulfilled. Likewise, the public interest

37 Urphy Vásquez Baca and Pedro Gamio Aita, 'Transición Energética Con Energías Renovables Para La Seguridad Energética En El Perú: Una Propuesta de Política Pública Resiliente Al Clima' [2018] Espacio y Desarrollo 195, 202, http://revistas.pucp.edu.pe/index.php/espacioydesarrollo/article/view/20184/20216.

38 Adolfo Céspedes Zavaleta, 'La Potestas Variandi de La Administración Pública En Los Contratos de Concesiones de Obras y Servicios Públicos' (2006) 0 Revista de Derecho Administrativo 203, 204, http://revistas.pucp.edu.pe/index.php/derechoadministrativo/article/view/16354/16760.

39 *Ibid.*

40 *Sentencia del Tribunal Constitucional de fecha 5 de julio de 2004 del proceso de amparo seguido por Juan Carlos Callegari Herazo contra el Ministerio de Defensa bajo el Expediente No 0090–2004-AA/TC* 1.

guides the design and creation of regulations and limits the exercise of political power and public administration.[41]

12.3.2 Timely Supply of Electricity

Protecting the public interest goes hand in hand with protecting the terms and conditions of the auction process and its participants. Thus, bid-awarded projects must be executed on time and within the proposed budget. This is the fundamental aspect of protecting the public interest since it involves State funds (public funds). Any deviation from the deadline or the planned budget would seriously affect the general public interest.[42]

In particular, the execution time of a project within the framework of an electricity supply concession is essential because it is related to the projected energy demand. If the project is executed after the deadline, the public interest may be affected to the extent that an unnecessary project is subsidised because the supply increases or the demand decreases. Thus, the RER regulation itself has established that MINEM has a legal impediment to modify the RER Supply Contracts concerning the end date of the contract[43] and the actual date of POC.[44]

12.3.3 The Attraction of Foreign Investors (by Long-Term Supply Contracts and Guaranteed Rates)

As mentioned, the mechanism established for promoting investment in this type of energy by the RER Law is an auction system, which

---

[41] Nicolás López Calera, 'El Interés Público: Entre La Ideología y El Derecho' (2010) 44 Anales de la Cátedra Francisco Suárez 123, 147, https://revistaseug.ugr.es/index.php/acfs/article/view/502/592.

[42] *Sentencia del Tribunal Constitucional de fecha 5 de julio de 2004 del proceso de amparo seguido por Juan Carlos Callegari Herazo contra el Ministerio de Defensa bajo el Expediente No. 0090–2004-AA/TC* (n 40).

[43] Republica del Peru Decreto Supremo No. 012–2011-EM (n 21) Article 1.13.B.

[44] *Ibid* Article 1.13.C.

aims to attract private investment in exchange for the State's long-term commitment to purchase the energy produced with RER.[45]

The generation promotion with RER under the auction/bidding mechanism implies granting benefits to the successful/awarded bidders. Among the most critical benefits given to investors by the RER Law are:

j) Priority of dispatch and purchase. The winner of an auction has priority for the daily dispatch of load and purchase of generated energy.[46]
k) Determination of applicable rates. Rates are determined at each auction; they are stable and guaranteed for a period of between 20 and 30 years.[47]
l) Connection priority. The successful/awarded bidders of an auction under the RER Law have priority to connect to the transmission and/or distribution SEIN systems.[48]

12.4 Analysis of Arbitration Cases Derived from RER Regime Projects

As mentioned above, the inclusion of arbitration as a tool to solve any controversy related to the execution of the RER Supply Contract is a guarantee of the RER Regime to promote private investment.,[49,50]

Since the awarding and execution of RER Regime projects has generated several controversies, it is relevant to analyse some of the most pertinent arbitration cases in which MINEM has taken the role of defendant.

[45] Vásquez Baca and Gamio Aita (n 37) 202.

[46] Republica del Peru (n 11) Article 5.

[47] *Ibid* Article 7.

[48] *Ibid* Article 8.

[49] *Laudo del arbitraje seguido entre Empresa de Generación Eléctrica Santa Lorenza SAC y Ministerio de Energía y Minas Caso Arbitral No 0672–2018-CCL* 17.

[50] Huapaya Tapia (n 34) 328–329.

12.4.1 Santa Lorenza S.A.C. ("Santa Lorenza") Against Minem

On 18 February 2014, Santa Lorenza and MINEM executed their RER Supply Contract. According to the latter, Santa Lorenza should have reached the POC by 31 December 2018, at the latest, as established in clauses 1.4.23 and 8.4 of the RER Supply Contract. Due to its impossibility to obtain the necessary permits for the project, Santa Lorenza requested an extension for the RER Supply Contract, and therefore, an extension of the POC date, which MINEM partially granted.[51]

Santa Lorenza's claim was based on the scope of the abovementioned clauses, specifically not reaching the POC on the maximum scheduled date (2 years from the expected date for POC proposed by Santa Lorenza) could generate, in any case, an automatic resolution of the RER Supply Contract.[52]

Santa Lorenza argued that the breach was caused by the delay of multiple government authorities in the issuance of relevant permits. Likewise, Santa Lorenza considered that the correct interpretation of the RER Supply Contract would be that its automatic resolution clause should not be applicable if the breach was due to a force majeure event (i.e. delay of a third entity or person).[53]

In addition, Santa Lorenza claimed that during the auction's "consultations and suggestions of the auction's bases" stage, the answers provided to the bidders related to the automatic resolution of the RER Supply Contract for the breach of the maximum scheduled date excluded force majeure cases were imprecise.[54]

The Tribunal concluded that, under a "literal interpretation" of Clauses 1.4.23 and 8.4, any breach of the maximum scheduled date to reach the POC phase generated an automatic resolution of the RER Supply Contract. In that sense, given that Santa Lorenza had access to the

---

51 *Laudo del arbitraje seguido entre Empresa de Generación Eléctrica Santa Lorenza S.A.C. y Ministerio de Energía y Minas. Caso Arbitral No. 0672–2018-CCL* (n 49).

52 *Ibid*.

53 *Ibid* 17–22.

54 *Ibid* 18.

text of the RER Supply Contract long before its execution, said company consciously assumed the mentioned risk.[55]

Further, the Tribunal noted that even though the answers provided during the auction's "consultations and suggestions" stage were not entirely clear, there was sufficient information to conclude that the bidders understood the risks they were assuming as bidders.[56]

12.4.2 Electro Zana S.A.C. ("Electro Zana") Against Minem

Similar to Santa Lorenza's case, on 18 February 2014, Electro Zana and MINEM executed their RER Supply Contract. According to the latter, Electro Zana should have reached the POC by 31 December 2018, at the latest, as established in clauses 1.4.23 and 8.4 of the RER Supply Contract. Due to its impossibility to obtain the necessary permits for the project, Electro Zana requested the RER Supply Contract's extension, and therefore, an extension of the POC date.[57]

According to Electro Zana, the delay of the competent governmental authorities in the issuance of the environmental impact assessment (EIA) and other relevant permits caused its delay to comply with reaching the POC on the maximum scheduled date.[58]

In this scenario, Electro Zana stated that they could not assume the consequences of any governmental authority's delay, given that they should be considered as their counterparty in the RER Supply Contract (as governmental entities are part of the State).[59]

Therefore, if MINEM decided to terminate the RER Supply Contract based on a delay caused by any governmental authority, it might be considered if such termination was caused by the party requesting it (i.e. MINEM). In this regard, common sense and good faith indicate that no

[55] *Ibid* 38–45.

[56] *Ibid* 45–47.

[57] *Laudo Arbitral del arbitraje seguido entre Electro Zaña SAC y Ministerio de Energía y Minas Caso Arbitral No 0677–2018-CCL* 20–22.

[58] *Ibid* 22.

[59] *Ibid* 25–28.

one can grant its counterparty the right to terminate an agreement for their own breaches.[60]

When analysing this case, the Tribunal started from the premise that the RER Supply Contract is considered a "regulated/administrative contract", which means that many of its clauses do not evidence the intention or will of the parties involved, as these clauses are a reproduction of provisions contained in the applicable laws (in this case, the RER Regime). Thus, the Tribunal stated that to determine the true meaning and scope of the RER Supply Contract clauses, the applicable laws and their ratio legis, including the dispositions that could not be contemplated in the agreement, must be considered.[61]

Therefore, given that the automatic resolution as a consequence for breaching the POC on the maximum scheduled date is a legal disposition also included in the RER Supply Contract, no governmental authority (including MINEM) could avoid such a result if said breach occurs.[62]

Finally, the Tribunal concluded that the RER Supply Contract was designed so that Electro Zana would assume the risk of breaching the POC on the maximum scheduled date because said date resulted from a two-year extension of the expected POC date provided by Electro Zana. In other words, Electro Zana had already been granted a two-year extension on such mentioned date.[63]

Bearing that in mind, the RER Regime provides that the breach of the POC on the maximum scheduled date (a result of a two-year extension of the expected date) generates an automatic resolution of the RER Supply Contract. The objective of such disposition on the RER Regime is to avoid any more extensions that would harm the schedule established for the reception of RER energy and, therefore, the national interest.[64]

---

60 *Ibid* 29–30.

61 *Ibid* 97–100.

62 *Ibid* 100.

63 *Ibid* 100–102.

64 *Ibid* 86, 109.

12.4.3 Empresa Generacion Hidraulica Selva Sociedad Anonima ("Egehissa") Against Minem

On 28 December 2011, Egehissa and Minem executed their RER Supply Contract. Egehissa required MINEM to extend the term of the RER Supply Contract and postpone the POC. MINEM acceded to Egehissa's request but demanded an increased faithful compliance with the execution of the works guarantee. Egehissa refused to deliver the increased guarantee to MINEM after said governmental authority approved an extension of the POC date; thus, MINEM terminated the RER Supply Contract.[65]

In this case, Egehissa argued the following: (i) it had already delivered a letter of guarantee when MINEM approved a previous extension of the POC date, and (ii) the cause that justified the mentioned extension was the same. Therefore, Egehissa considered an additional guarantee was unreasonable.[66]

Egehissa based its claim on the fact that the RER Supply Contract did not establish that every extension of the POC date implied the presentation of a new letter of guarantee or an increased guarantee, even more, when the delays were generated by the same events caused by governmental authorities.[67]

Regarding the RER Supply Contract termination by MINEM, Egehissa also stated that MINEM did not follow the contractual resolution procedure, depriving it of its right of defence.[68]

The tribunal concluded that Egehissa did not have to present an additional or increased letter of guarantee for the latest approved extension. Given that said guarantees already covered MINEM in the event Egehissa could not comply with the work execution schedule even after

65 *Laudo del arbitraje seguido entre Empresa Generación Hidráulica Selva Sociedad Anónima - Egehissa y Ministerio de Energía y Minas Caso Arbitral No 0087–2019-CCL* 1, 6–8.

66 *Ibid* 8–11.

67 *Ibid* 11–12.

68 *Ibid* 12–15.

the extension, with only one guarantee Egehissa was covering such risk, and there was no justification for the presentation of another guarantee.[69]

Therefore, the Tribunal, as MINEM was not able to prove the reasons that justified the request of an additional letter of guarantee, concluded MINEM's request was exorbitant, abusive, manifestly unreasonable, and out of common sense.[70]

Finally, the Tribunal stated that if MINEM should terminate the RER Supply Contract, it shall follow the procedure established in said agreement. Hence, the termination of the Egehissa RER Supply Contract was ineffective, and the contractual relationship was still binding.[71]

12.4.4 Empresa De Generacion Electrica Colca S.A. ("Egecolca") Against Minem

On 18 February 2014, Egecolca and MINEM executed their RER Supply Contract. Egecolca's case was similar to Santa Lorenza and Electro Zana's cases. Due to a delay of governmental authorities in the issuance of relevant permits, Egecolca could not comply with the POC on the maximum scheduled date.[72]

Even though Egecolca requested to MINEM the modification of the RER Supply Contract's POC maximum scheduled date and termination date, MINEM automatically terminated the referred contract due to the breach of the mentioned maximum scheduled date.[73]

As with Santa Lorenza and Electro Zana, Egecolca based its allegations on the fact that the breach of the POC on the maximum scheduled date was caused by events not attributable to it.[74]

The Tribunal started from the premise that the RER Supply Contract is considered an "administrative contract" when analysing this case.

---

69 *Ibid* 23–26.

70 *Ibid* 57–64.

71 *Ibid* 79–86.

72 Centro de Arbitraje de la Cámara de Comercio de Lima, 'Laudo Arbitral de Derecho Del Arbitraje Seguido Entre Empresa de Generación Eléctrica Colca S.A.C. y Ministerio de Energía y Minas. Caso Arbitral N° 0651–2018-CCL' 1, 23–24.

73 *Ibid* 26.

74 *Ibid* 27.

Consequently, its characteristics and purposes are determined by the national interest and public necessity to have an adequate energy system. Therefore, any possible modification that arises during the execution of the agreement affects the parties involved and the public as electricity consumers.[75]

In that sense, considering what the parties agreed when executing the RER Supply Contract, the Tribunal concluded that any delay or breach of the POC on the maximum scheduled date, for whatever reason, would be unjustified and would lead to the immediate and automatic resolution of the RER Supply Contract.[76]

12.4.5 Recurrent Issues on the Analysed Arbitration Cases

According to the referred cases of the previous sections, it is possible to conclude that all the arbitration cases relate to different requests to extend the POC due to various reasons; and a request to determine if MINEM can demand increased guarantees in the event that it grants an extension of the POC.

In the following paragraphs, we will analyse if the arbitration awards ratify the critical aspects of the RER Regime detailed; and if some improvements can be made either to the auction bases, the RER Contract model, or the RER Regime.

m) Requests to delay the POC on the maximum scheduled date

The Santa Lorenza, Electro Zana, and Colca arbitration cases relate to different requests of the awarded bidders and parties to the specific RER Supply Contracts to extend/delay the POC maximum scheduled date of the RER awarded projects. According to some clauses of the RER Supply Contracts, these delay requests are based either on force majeure issues or delays attributed to MINEM.

---

[75] *Ibid* 52–53, 95.

[76] *Ibid* 53–57.

- Force majeure issues

In the Santa Lorenza case, Santa Lorenza alleged that it could not reach the POC due to the delay of multiple government authorities in the issuance of relevant permits. Santa Lorenza characterised the latter as an event of force majeure, considering that neither of the RER Supply Contract parties was responsible for such delay.[77]

Likewise, Egecolca argued that due to the delay of governmental authorities in issuing relevant permits (events not imputable to Egecolca), it could not comply with the POC on the maximum scheduled date.[78]

On both arbitration cases, the arbitral tribunals concluded that the spirit and correct interpretation of clauses 1.4.23 and 8.24 of the RER Supply Contracts was that: (i) the consequence of not reaching the POC on the maximum scheduled date implied an immediate and automatic resolution of the mentioned agreement[79]; and, (ii) since the RER Supply Contracts are administrative/regulated agreements, any amendment or modification should consider the national interest and public necessity of the RER Regime, as these amendments not only have an impact among the parties of the RER Supply Contract but most importantly, the public as consumers.[80]

Based on the arbitral tribunal considerations and arguments, it is possible to conclude that the RER Supply Contract has the nature of an administrative contract, regulated by the RER Regime, which aims to materialise the State policy to promote the use of RER in electricity generation. Given that the RER Regime is also a promotional one (access to payment of fees by the State to the awarded bidders), the RER

[77] *Laudo del arbitraje seguido entre Empresa de Generación Eléctrica Santa Lorenza S.A.C. y Ministerio de Energía y Minas. Caso Arbitral No. 0672–2018-CCL* (n 49).

[78] Centro de Arbitraje de la Cámara de Comercio de Lima (n 72) 27.

[79] *Laudo del arbitraje seguido entre Empresa de Generación Eléctrica Santa Lorenza S.A.C. y Ministerio de Energía y Minas. Caso Arbitral No. 0672–2018-CCL* (n 49).

[80] Centro de Arbitraje de la Cámara de Comercio de Lima (n 72) 52–53, 95.

Regime must have a clearly defined time horizon of the validity of these promotional incentives.[81]

Likewise, any interpretation of the RER Supply Contract that modifies its administrative/regulated nature to a civil contract must be rejected. Thus, it must be considered that the purpose of the State in these contracts is to pursue the Regime's public interest. Consequently, the State seeks to generate electricity with RER, safeguard the use of public goods[82] (e.g. water, air) and guarantee the energy security of users in general. In this way, the Peruvian State must guarantee the legal protection of the RER Regime and not allow the regime's purpose to be harmed.[83]

Based on these considerations, the POC cannot exceed the referential POC date by more than two years. This limitation in the RER Supply Contracts replicates provisions of the RER Regime, without prejudice to other provisions of this contract that might be the product of the will of the parties[84] (i.e. offer of the awarded bidder, payment of tariffs and so on).

- Minem's attributed delay on granting different permits

As stated by Electro Zana in its arbitration case, it was unable to reach the POC on the maximum scheduled date due to the delay of the competent governmental authorities in the issuance of relevant permits.[85] The company also indicated that it could not assume any delay caused by any governmental authority, as these authorities should be considered as its counterparties in the RER Supply Contract (as part of the Peruvian State).[86]

---

[81] *Laudo del arbitraje seguido entre Empresa de Generación Eléctrica Santa Lorenza S.A.C. y Ministerio de Energía y Minas. Caso Arbitral No. 0672–2018-CCL* (n 49) 50.

[82] *Ibid* 53.

[83] Centro de Arbitraje de la Cámara de Comercio de Lima (n 72) 95.

[84] *Laudo del arbitraje seguido entre Empresa de Generación Eléctrica Santa Lorenza S.A.C. y Ministerio de Energía y Minas. Caso Arbitral No. 0672–2018-CCL* (n 49) 54.

[85] *Laudo Arbitral del arbitraje seguido entre Electro Zaña S.A.C. y Ministerio de Energía y Minas. Caso Arbitral No. 0677–2018-CCL* (n 57) 22.

[86] *Ibid* 25–28.

Thus, according to Electro Zana, the RER Supply Contract's termination by MINEM should be considered illegal, as common sense and good faith principles state that no one can grant its counterparty (MINEM and governmental authorities) the right to terminate an agreement for their own breaches.[87]

Similar to Santa Lorenza and Egocolca's cases, the arbitral tribunal of Electro Zana's case considered the RER Supply Contract as a regulated/administrative agreement; therefore, some of its clauses reproduce provisions contained in the RER Regime. According to the tribunal, when interpreting the RER Supply Contract, the RER Regime and its *ratio legis* shall be considered to establish the contract's meaning and scope.[88]

Based on the previous, it is possible to conclude that an automatic resolution of the RER Supply Contract directly results from not reaching the POC on the maximum scheduled date. This clause of the RER Supply Contract is a legal disposition established in the RER Regime; thus, neither MINEM nor any governmental authority is legally entitled to extend or delay the maximum scheduled date for POC.[89]

In addition, the Tribunal stated that: (i) Electro Zana assumed the risk of POC, as the RER Supply Contract and the RER Regime clearly establish the consequences for such breach, as well as the risk to obtain in a timely manner all the permits necessary for the execution of the awarded project; and (ii) Electro Zana had already been granted a two-year extension to reach POC.

According to the Arbitral Tribunal in the case of Electro Zana, the clause that establishes the automatic termination of the RER Contract is a resolutory/dissolving condition, and that is how clause 8.4 of the RER Supply Contract shall be interpreted. To sustain this affirmation, the Tribunal considered that MINEM does not have the legal authority to extend the term of the RER Supply Contract if the maximum scheduled date for POC[90] was breached, as established in the RER Regime.

---

[87] *Ibid* 29–30.

[88] *Ibid* 97–100.

[89] *Ibid* 100.

[90] *Ibid*.

In this sense, the resolutory/dissolving condition must be understood to be fulfilled if the POC was not reached on the maximum date allowed. Thus, MINEM does not need to make any communication or carry out any activity in this regard since the consequence of clause 8.4 of the RER Supply Contract occurs automatically.[91]

n) Consultation and suggestions to the Bases stage

In Santa Lorenza's case, the company argued that the answers provided to the bidders during the auction´s "consultations and suggestions of the auction's bases" stage (related to the automatic resolution of the RER Supply Contract and MINEMs responsibility if permits were not granted) were vague and nonspecific.[92]

Despite Santa Lorenza's claims, the Arbitral Tribunal considered that the bidders of the particular auction (third bid) were conscious of the risk of automatic termination of the RER Supply Contract if the POC was not reached on the maximum scheduled date. In addition, the Tribunal considered that the phrase "*por cualquier motivo*" (for any circumstance) in clause 8.4 of the RER Supply Contract includes any force majeure event.[93]

The Tribunal also reviewed[94] the consultations made to the auction bases to verify the answers provided to the bidders during the "consultations and suggestions of the auction's bases" stage. Thus, the Tribunal analysed in particular the questions related to the automatic resolution of the RER Supply Agreement and its application in events of force majeure.[95]

The Tribunal was emphatic when it declared that it did not share Santa Lorenza's claim on the vagueness of the answers provided during

91 *Ibid* 103.

92 *Laudo del arbitraje seguido entre Empresa de Generación Eléctrica Santa Lorenza S.A.C. y Ministerio de Energía y Minas. Caso Arbitral No. 0672–2018-CCL* (n 49) 18.

93 *Ibid* 45–46.

94 *Ibid*.

95 Organismo Supervisor de la Inversión en Energía y Minería - OSINERGMIN, 'Circular No. 01 de La Tercera Subasta de Suministro de Electricidad Con Recursos Energéticos Renovables' (2013) 10, 15, https://www.osinergmin.gob.pe/seccion/centro_documental/energias-renovables/Subastas/TerceraSubasta/tercerasubasta_Circular1.pdf.

the consultation stage. According to the Tribunal, the responses of the consults were evident, as they emphasised that the automatic resolution of the RER Supply Contract contained in clause 8.4 included events of force majeure.[96]

In this sense and in consideration of the Arbitral Tribunal's position, it is possible to indicate that the auction process ensures that each bidder has clear, sufficient and complete information that allows them: (i) to propose the project with which they will participate in the auction; and (ii) make all inquiries and observations to the bases during the auction process and thus adequately define their participation and offer.[97]

Consequently, the auction process, and in particular the 'consultations and suggestions of the auction's bases' stage, allows the bid participants to know what risks have been assigned to them and that this assignment of risks, in specific matters (such as automatic termination due to not reaching the POC) it is not alterable by legal mandate of the RER Regime.

o) Additional requests by Minem to the RER Contract

As mentioned in Egehissa's arbitral dispute, MINEM requested the company's increased faithful compliance with the execution of works guarantee to accede and postpone the POC. As Egehissa did not deliver the increased guarantee, MINEM terminated the RER Supply Contract.[98]

The tribunal concluded that: (i) Egehissa did not have to present an additional or increased letter of guarantee; (ii) the guarantees already granted in MINEMs favour already covered the latter in the event Egehissa could not comply with the work execution schedule and (iii) there was no justification for the presentation of another guarantee.[99]

---

96 *Laudo del arbitraje seguido entre Empresa de Generación Eléctrica Santa Lorenza S.A.C. y Ministerio de Energía y Minas. Caso Arbitral No. 0672–2018-CCL* (n 49) 47.

97 De la Piedra Calle (n 19) 220.

98 *Laudo del arbitraje seguido entre Empresa Generación Hidráulica Selva Sociedad Anónima - Egehissa y Ministerio de Energía y Minas. Caso Arbitral No. 0087–2019-CCL* (n 65) 6–8.

99 *Ibid* 23–26.

Based on this case's arbitration award, we can conclude that MINEM cannot increase the amount of the guarantees for any cause without proper justification. Otherwise, such requests will be deemed exorbitant, abusive, manifestly unreasonable and out of common sense.[100] In addition, MINEM cannot terminate a RER Supply Contract without following the procedure established in such agreement (if the termination clause is not deemed as a resolutory/dissolving condition).[101]

12.4.6 Proposed Measures to Decrease Litigation Related to RER Supply Contracts

As previously analysed, most of the arbitration cases refer to: (i) clause 8.4 of the RER Supply Contracts; (ii) the conditions for its immediate application and (iii) if force majeure events or the delay in obtaining relevant permits means the nonapplication of such clause.

The analysed arbitration awards clearly delimitate the meaning and purpose of such a clause. However, some interpretation issues appear when analysing the clause's text and its conjunct application with other terms and conditions of the RER Supply Contract.

p) Redrafting of clause 8.4 of the RER Supply Contract

In the third and fourth bids, Clause 8.4 of the RER Supply Contract established:

> If as of December 31, 2018, for any reason, the Commercial Operation Commissioning [POC] has not taken place of the RER Generation Project which is the object of this Contract, it will be resolved of full right, and the Guarantee of Faithful Compliance will be executed.
>
> The resolution of full right to which the previous paragraph refers, will not affect the enforcement of the final generation concessions and that of

100 *Ibid* 57–64.

101 *Ibid* 79–86.

transmission that have been granted according to the Law of Electrical Concessions and Bylaws of Law of Electrical Concessions.[102]

While the redaction and drafting of this clause seem clear, the analysed arbitration cases show that a more precise language is needed. Thus, we consider and recommend that the phrase "for any reason" is insufficient and should also clearly refer to (i) force majeure events and (ii) delays of governmental authorities in granting necessary permits.

With this amendment, we consider that (i) there will be no room for further and erroneous interpretations by any awarded bidder and (ii) consultations and suggestions to modify the clause will be reduced, as it will consider all aspects (force majeure and delays to obtain permits) that have been litigious.

q) Precision to MINEMs "contribution" to acquire or obtain necessary permits

In the Electro Zana arbitration case, it stated that the cause for the POC not taking place was the delay of the competent governmental authorities in the issuance of relevant permits.[103]

In the third bid, Clause 4.3 of the RER Supply Contract establishes:

> 4.3 The Ministry will impose the rights of way which are required according to what is established in the Applicable Laws, but will not cover the costs incurred to obtain such rights of way.

---

[102] The Spanish text of Clause 8.4 for the Third and Fourth Bids estates:

Si al 31 de diciembre de 2018, por cualquier motivo, no se ha concretado la Puesta en Operación Comercial del Proyecto de Generación RER materia del presente contrato, éste quedará resuelto de pleno derecho, ejecutándose la Garantía de Fiel Cumplimiento.

La Sociedad Concesionaria mantendrá los derechos sobre la Concesión Definitiva de Generación que le hubiere sido otorgada, de acuerdo a los términos de la LCE y el RLCE.

Third bid, p. 65, https://www.osinergmin.gob.pe/seccion/centro_documental/energias-renovables/Subastas/TerceraSubasta/tercerasubasta_Bases_3ra_subasta_RER_Consolidadas.pdf.

Fourth bid, p. 71, https://www.osinergmin.gob.pe/seccion/centro_documental/energias-renovables/Subastas/Bases_Consolidadas_4taSubasta.pdf.

[103] *Laudo Arbitral del arbitraje seguido entre Electro Zaña S.A.C. y Ministerio de Energía y Minas. Caso Arbitral No. 0677–2018-CCL* (n 57) 22.

Likewise, if **it is required by the Concessionaire Society, the Ministry** will make every effort so that it will have access to installations of third parties and, **will contribute to the acquiring of permits, licenses, authorizations, concessions, rights of way, rights of use and similar in case these are not granted by the competent Government Authority in the due time**, notwithstanding the fact that the requirements and procedures necessary according to the Applicable Laws have been complied with".[104] (emphasis added).

An amendment to this clause was made for the fourth bid, as it established that:

4.3 The Ministry will impose the rights of way which are required according to what is established in the Applicable Laws, but will not cover the costs incurred to obtain such rights of way.

Likewise, if it is required by the Concessionaire Society, the Ministry will make every effort so that it will have access to installations of third parties and, will contribute in the acquiring of permits, licenses, authorizations, concessions, rights of way, rights of use and similar in case these are not granted by the competent Government Authority in the due time, notwithstanding the fact that the requirements and procedures necessary according to the Applicable Laws have been complied with. **This, in no case will imply the assumption of any type of responsibility of the Grantor and cannot be considered as a condition for the compliance of the obligations of the Concessionaire Society**".[105] (emphasis added).

---

[104] The Spanish text of Clause 4.3 for the Fourth Bid estates:

4.3 El Ministerio impondrá las servidumbres que sean requeridas de acuerdo a lo establecido en las Leyes Aplicables, pero no asumirá los costos incurridos para obtener dichas servidumbres.

Asimismo, de ser requerido por la Sociedad Concesionaria, el Ministerio hará sus mejores esfuerzos para que aquella acceda a instalaciones de terceros y coadyuvará en la obtención de permisos, licencias, autorizciones, concesiones, servidumbres, derechos de uso y similares, en caso estos no fueran otorgados por la Autoridad Gubernamental competente en el tiempo debido, a pesar de haberse cumplido los requisitos y trámites exigidos por las Leyes Aplicables. Ello, en modo alguno implicará la asunción de algún tipo de responsabilidad en el Concedente y no podrá ser considerado como condición para el cumplimiento de las obligaciones de la Sociedad Concesionaria.

Organismo Supervisor de la Inversión en Energía y Minería—OSINERGMIN, 'Bases Consolidadas Para La Tercera Subasta de Suministro de Electricidad Con Recursos Energéticos Renovables' (2013) 60, https://www.osinergmin.gob.pe/seccion/centro_documental/energias-renovables/Subastas/TerceraSubasta/tercerasubasta_Bases_3ra_subasta_RER_Consolidadas.pdf.

[105] The Spanish text of Clause 4.3 for the Fourth Bid estates:

Even though an additional sentence has been added to clause 4.3 to limit MINEMs responsibility, we consider that the phrase "the Ministry will contribute in the acquiring of" is still misleading. Such "contribution" obligation has no legal grounds (administrative competence) or consequences, either in the RER Supply Contract or in the RER Regime. Similarly, Egecolca's arbitration award states that MINEM cannot be held responsible for other governmental authorities' acts, as it was not responsible for (i) the procedures to grant such permits, nor (ii) caused the delays.[106]

In addition, the "contribution" obligation crashes with other relevant clauses of the RER Supply Contract (i.e. clauses 3.2, 3.3, 4.1, 4.2, 4.3 and following) that evidence the allocation of design risk, structuring and obtaining the relevant permits on the awarded bidders.[107]

12.5 Conclusions

The RER Regime contains special and specific regulations that seek to guarantee the timely and efficient supply of electricity generated with RER to the population, establishing specific incentives for investors awarded in an auction procedure.

4.3 El Ministerio impondrá las servidumbres que sean requeridas de acuerdo a lo establecido en las Leyes Aplicables, pero no asumirá los costos incurridos para obtener dichas servidumbres.

Asimismo, de ser requerido por la Sociedad Concesionaria, el Ministerio hará sus mejores esfuerzos para que aquella acceda a instalaciones de terceros y coadyuvará en la obtención de permisos, licencias, autorizciones, concesiones, servidumbres, derechos de uso y similares, en caso estos no fueran otorgados por la Autoridad Gubernamental competente en el tiempo debido, a pesar de haberse cumplido los requisitos y trámites exigidos por las Leyes Aplicables. Ello, en modo alguno implicará la asunción de algún tipo de responsabilidad en el Concedente y no podrá ser considerado como condición para el cumplimiento de las obligaciones de la Sociedad Concesionaria.

Organismo Supervisor de la Inversión en Energía y Minería—OSINERGMIN, 'Bases Consolidadas Para La Cuarta Subasta de Suministro de Electricidad Con Recursos Energéticos Renovables' (2015) 63–64, https://www.osinergmin.gob.pe/seccion/centro_documental/energias-renovables/Subastas/Bases_Consolidadas_4taSubasta.pdf.

[106] Centro de Arbitraje de la Cámara de Comercio de Lima (n 72) 106.

[107] Organismo Supervisor de la Inversión en Energía y Minería—OSINERGMIN, 'Bases Consolidadas Para La Tercera Subasta de Suministro de Electricidad Con Recursos Energéticos Renovables' (n 104) 59–60.

The auction procedure has mechanisms that guarantee the transparency of the requirements of the generation project with RER. In this sense, any bidder has sufficient and timely information regarding the project that it wishes to implement, as well as the technical and legal requirements that they must comply with in order to be awarded.

The bidders that participate in the auctions for the supply of electricity generated with RER design the specific projects to offer the amount of energy required and its price. The best bidder is the successful bidder and signs a supply contract with MINEM, which guarantees the purchase of all the electricity generated for a period of up to 20 years at the price proposed by the successful bidder. In return, this supply contract assigns to the awarded bidder/concessionaire the risks of design, obtaining authorisations, permits and licences, construction, and operation and maintenance of the project.

RER Supply Contracts are administrative contracts regulated by their own provisions and, in the event of gaps, by the RER Law and its regulations or RER Regime, and, additionally, by the Peruvian Civil Code. Additionally, certain principles of the private investment promotion regime (like the use of arbitration for disputes) and respect for fundamental rights are applicable to said contracts, always within the framework established by the RER Supply Contract, the Consolidated Bidding Bases and the RER Regime.

Minem is not legally nor administrative empowered to change the risk allocation of the RER Supply Contracts, as established in the RER Regime and the analysed arbitration awards. In addition, MINEM is legally prevented from agreeing to modify the RER Supply Contract's Commercial Operation Commissioning and End Date, as any amendment to these sections implies the violation of specific provisions contained in articles 1.13.B and 1.13.C of the RER Regulation. By the provisions of the Regulation above, the date of Commercial Operation Commissioning and the End Date of the RER Supply Contract are not modifiable for any reason. Their breach implies the automatic termination of the RER Supply Contract and the execution of the guarantee of faithful compliance.

The public interest of the RER Regime cannot have a biased analysis or that only focuses on the convenience of carrying out a particular project; on the contrary, it must have a more inclusive and reasonable analysis in the light of the legislation for the promotion of RER projects. In this regard, it is in the public interest and in accordance with the RER Regime that the commercial terms and term awarded for the promotion and development of these projects are not altered.

The arbitration awards reviewed confirm that the requests for modification of the RER Supply Contracts have limitations expressly contained in the RER Supply Contract and in the RER Regime itself. Provisions 1.13.B and 11.13.C of the RER Regulation and the provisions of clause 8.4 of the RER Supply Contracts are consistent with the national interest and public need for the timely and efficient use of the electricity supply generated with RER that Peru seeks to promote. However, it is necessary to vary/modify some sections of the RER Supply Contract to reflect the mentioned provisions of the RER Regulation and decrease related litigation.

13

Energy Transition in Latin American Countries and the Conflicts Associated with Such Transitions

Claudia Fonseca Jaramillo

13.1 The Approach to Energy Transition in Latin America

A correct approach to energy transition in any country, but especially in countries with many unmet state needs,[1] is to achieve a good balance between energy production, sources of such production, reduction of environmental impacts, the guarantee of permanent supply and a guarantee of financing and payment. This is of the utmost importance because when talking about energy transition, many people believe

[1] Such as the lack of electrical power supply in many regions or the non-constant supply in other regions.

C. F. Jaramillo (✉)
Garrido Fonseca Abogados, Bogotá, Colombia
e-mail: claudia.fonseca@garridofonseca.com

G. Wood and J. F. Neira-Castro, *From Fossil Fuels to Low Carbon Energy Transition*, Energy, Climate and the Environment,
https://doi.org/10.1007/978-3-031-00299-1_13

that it is simply just a change from non-renewable to renewable energies. However, there is a lack of awareness concerning the concept of energy security, which includes not only electricity production but also environmentally, socially and economically sustainable sources of such production.

In developing countries, where the energy matrix depends mostly on fossil sources, the manner in which the energy transition is conducted must, on one hand be very well coordinated so as to ensure a constant supply, while on the other remain economically and financially viable in order to meet the objectives of the Paris Agreement (2015), which aims to '*decarbonise the economy by migrating from fossil energies to non-conventional sources of renewable energies*'.

If an optimal balance between supply, sources and financing is not achieved, developing countries will be unable to meet their expected decarbonisation agenda. There will be no financing sources to guarantee their long-term energy supply since multilateral organisations have begun to close credit lines for the exploration and exploitation of non-renewable energy sources. An example of this is the closure of all credit lines by the World Bank for the financing of coal, oil and gas exploration and exploitation projects.[2]

For a country such as Colombia, for example, it is estimated that, in the medium term, renewable energies could replace the installed electric power capacity originating from coal sources. However, the installed capacity originating from hydrocarbon and gas sources cannot be replaced by renewable sources, at least in the medium term, and gas will have to be used for many more years. In this context, consideration should be given to the prospect that the matrix will be dependent on gas for decades to come. Given this, Latin American countries should be thinking about extracting the largest amount of hydrocarbon resources in the shortest possible time, and use the economic resources of such extraction to guarantee a continuous and reliable supply in the long term. Moreover, this would also meet the standards of the Paris Agreement,

[2] Acosta, Amylkar. "THE DECLINE OF COAL: A SERIOUS CHALLENGE FOR COLOMBIA" (2021) Article published in El Tiempo newspaper, Public Reason Section, https://www.eltiempo.com/economia/el-ocaso-del-carbon-es-un-serio-reto-para-colombia-568428.

which in the case of Colombia, implies a 20% reduction of its Greenhouse Gas (GHG) emissions by 2030, based on the GHG inventory for the year 2010.

Additionally, in countries such as Colombia which, despite having one of the cleanest energy matrices on the planet thanks to its water resources and hydropower installed capacity, also face significant backup problems in El Niño years, which may lead to large energy shortages. This then needs to be supplied from other sources that must be available at all times to avoid situations of high energy supply shortages, such as the one that occurred back in 1992. This incident cost Colombia many millions of dollars of investment in thermoelectric plants to avoid a 'power outage' like the one sustained in the nineties. This in turn led to an increase in the electricity rate with the so-called 'cargo por confiabilidad' (reliability charge).

In 2018, energy generation using renewable sources accounted for 58% of the total energy generated in Latin America and the Caribbean. Much of this energy was generated by hydroelectric plants. However, today the construction of hydroelectric plants has a questionable role due to the following impacts; water use, climate change, impact on vulnerable ecosystems,[3] decrease in river flows and water quality due to the increase of sediments. This situation, resulting from large-scale hydroelectric projects, becomes even more complex if we add the prolonged and intense droughts derived from the El Niño phenomenon, making countries liable to a high degree of future vulnerability if their energy structure is not diversified. Therefore, for the above reasons, the contribution of hydroelectric energy production should be replaced with other lower impact sources, such as wind energy and solar photovoltaic energy, all of which the region also has substantial as yet untapped potential.

Brazil for their part, with the oil exploration of the 'Pré-Sal', and Argentina, with the extraction of shale gas in the 'Vaca Muerta' fields, are delaying significant changes to their energy matrices. For example, 70% of the energy supply in Argentina still comes from thermal sources (fossil

[3] Claver-Carone, Mauricio, "ENERGY TRANSITION IN LATIN AMERICA AND THE CARIBBEAN". (2021), "Energy transition: a legacy for Colombia's present and future", https://www.minenergia.gov.co/libro-transicion-energetica.

fuels), while in Brazil, import of petroleum products, such as naphtha and diesel oil are growing and remain pillars of its energy supply chain,.[4]

This being the case, Argentina, Chile and Brazil have not increased investment in renewable energy projects in recent years, as shown by McCrone et al. (2019) who indicated that in 2019, Brazil decreased its investment in renewable energy by 44% and Chile decreased investment in clean energy projects by 38% in the previous five years.[5]

Moreover, in Latin America and the Caribbean, there are still many inhabitants without access to electricity. It is estimated that around 4.5 million people do not have access to an energy supply and 60% of the companies in the region still experience interruptions in providing electricity according to the Inter-American Development Bank (IDB) in its publication on energy access in 2020.[6]

Taking into account the above scenario on energy generation and energy supply coverage in Latin America and the Caribbean, to date, each country should have planned how to make the transition and which national resources should count towards enabling such a transition. The unused resources of today utilised as a driving force to achieve a long-term sustainable matrix will be resources that cannot be used in the future, either because they will be banned in the future, will become obsolete or simply will not be available at a later stage. Notably, this is without considering the great demand for 'clean' technology, which is already evidenced throughout the world, and which will most likely leave developing countries yet again at the back of the queue.

An example of the above is what is currently happening in China. With its commitment to reach carbon neutrality by 2060 and with, at present, huge coal consumption, China must invest eye-watering amounts of money to make the desired transition. This is arguably why China controls about 80% of the world's exports of solar panels, produces most of the materials required to manufacture vehicles and grid storage equipment, is a leader in electric vehicle production and has

---

[4] Egler, Claudio A. G. "Crisis and Energy transition in South America" (2020), https://www.researchgate.net/publication/345767298_Crisis_y_transicion_energetica_en_America_del_Sur/link/5fad390892851cf7dd13a048/download.

[5] Ibidem.

[6] Claver-Carone, Mauricio, idem.

more ambitious production targets than any other nations. All of this dovetails with China's national strategy to reduce dependence on fossil fuel imports—now 75% for oil—and to curb the politically crippling impacts of pollution.[7]

Undoubtedly, China will be followed by other major powers in terms of demand for equipment and technologies. If the countries of Latin America and the Caribbean do not have a clear roadmap to achieve the energy transition to which they have committed, they will fall behind. Moreover, they will be failing to meet their obligations and thus contributing to global warming.

13.2 Timeline for Transition

For this reason, there should already be a clear roadmap in each Latin American country for changes in national energy matrices. This does not need to be simply a shift from fossil fuels (high carbon) to renewable energy sources (low carbon). It could also, for example, include the conversion from coal to natural gas, alongside incorporation of additional kilowatts generated from renewable sources including wind power and solar power. This would have the advantage of providing clarity on the long-term supply of gas and hydrocarbons, while clean energies begin their expansion, thus providing each nation with certainty on the source of backup energy supply.

To give an idea of the dimensions, Colombia, for example, has installed a total of 17,312 Megawatts (MW), or approximately 17.3 gigawatts (GW) of total electricity generation capacity, yet just 19.5 MW of this total is provided by wind energy. With such numbers, we can visualise the size of Colombia's performance in terms of the scale of global

[7] Yergin, Daniel, "THE ENERGY TRANSITION OF COLOMBIA". (2021), "Energy transition: a legacy for Colombia's present and future", https://www.minenergia.gov.co/libro-transicion-energetica.

wind projects, considering that 620 GW of wind energy was generated in the world at the end of 2019.[8]

These days, the installed electric power capacity originating from coal sources in Colombia is about 1,612 MW. Looking at the current wind energy projects registered with UPME (Mining and Energy Planning Unit – Colombia), at least on paper, this should be able to replace the sources originating from coal. The same cannot be said about the energy sources originating from gas. With an installed capacity of 2,129 MW, that has been growing steadily, for example, around two-thirds (64.5%) of Colombian households today have access to this energy source.[9] All this, without going into a detailed analysis of, among other things, the geographical location of sources and transmission capacity, can affect the availability, efficiency and effectiveness of wind power generation. This led us to our first conclusion that, under the current circumstances, Colombia will continue its dependence on gas for the foreseeable future.

The energy produced in Colombia (taking electric power as a source of energy) mainly depends on water and fossil fuels. If, as is foreseen, there is a transition away from these fossil fuels towards cleaner energy sources, this has been argued to inevitably lead to an increase in electric power consumption via electrification. Thus, it could further be argued that wind energy will most probably not be a replacement power source, but rather a complement to meet the increased demand. Indeed, this has already been shown elsewhere in the world.

Therefore, under this scenario, Colombia and the other Latin American and Caribbean countries should be guaranteeing a long-term and continuous supply of natural gas, to ensure their energy security. Otherwise, they may sustain power outages just like the one that took place in Colombia back in 1992. This is especially pertinent in Colombia as gas reserves are, at present, limited and will only suffice for a few years. For this reason, it is essential to continue searching for new reserves

[8] BP web page, "Statistical Review of World Energy - Renewable energy", https://www.bp.com/en/global/corporate/energy-economics/statistical-review-of-world-energy/renewable-energy.html#wind-energy.

[9] Naturgas Web Page, "Indicators 2019", (2019), https://www.naturgas.com.co/documentos/2019/Indicadores2019.pdf.

in order to extend the country's energy self-sufficiency horizon, while making progress in the transition process.

The Colombian National Hydrocarbons Agency has been timidly stimulating the exploration of new conventional and unconventional continental and offshore fields and reservoirs, as it is believed that there is high gas prospectivity. However, investments are not forthcoming, and some sectors have been very reluctant to support and back the exploration of unconventional deposits even though this could be the solution to the longed-for energy transition in Colombia. This is comparable with what is happening in Argentina and Brazil with, 'contrario sensu'. Here, state strategy has placed the emphasis on investing in non-renewable energies, mainly for thermal energy generation, and thus neglecting investment in renewable energies.

In view of the above, some Latin American countries are nonetheless in a desirable situation in terms of transition. This is the case of Costa Rica, where the production of energy from renewable electricity sources has been at around 98%[10] for several years now. However, in countries such as Brazil, Chile or Argentina, a large part of the energy matrix remains still based on non-renewable resources. In particular, Brazil and Argentina will delay the transition until their non-renewable potential begins to decrease. There are also intermediate countries such as Colombia where the transition is lively. In such cases, then, fundamental decisions must be made regarding the speed of transition and the ways to finance it, considering the country has a medium potential for gas and oil, and other natural resources that can still be sought and exploited; otherwise they will remain buried forever.

Taking into account the above scenario, it is important to bear in mind that each country in Latin America and the Caribbean face their own particular situations. But all of them are faced with energy transitions and are in need a clear path to achieve this. This plan should enable them to achieve: a balance between the resources present in their countries, the exploitation rate of such resources, remain committed to a global track record in reducing carbon emissions, maintain constant

---

[10] Alonso, Judith, "Energy Transition: Latin America in the race" (2019), https://www.dw.com/es/transici%C3%B3n-energ%C3%A9tica-am%C3%A9rica-latina-en-la-carrera/a-48310900.

sources of financing and, most importantly, have future access to a stable and constant supply of energy, so that they do not fall at some point due to lack of energy system planning.

In this scenario, each country needs to be clear that, if there are delays, non-compliance or incidents in the transition plan, all blame will fall on the government, and both the state and state entities may end up being sued for their current decisions by companies that made investments on this basis. This could also result in companies and investors becoming frustrated in the future or unable to make the planned investments due to situations that each government in the region should have foreseen from the beginning yet did not do so. Thus, the long-term path for the energy transition must be unambiguous and structured so that investors have a clear set of rules in place which are aligned with state objectives.

In order to avoid conflicts that could lead to future arbitration claims in energy matters, each Latin American country should be clear on at least the following variables:

1. The international commitments of each country in environmental matters, especially those related to the Paris Agreement
2. The actual date on which each country committed to achieving the objectives of the Paris Agreement (2015), which aims to *'decarbonise the economy by migrating from fossil energies to non-conventional sources of renewable energy sources'*.
3. A true and verifiable schedule to make the energy transition, with clarity on which resources are to be used during each stage, and when, for example coal, will no longer be used and how much gas will be necessary in the future to replace coal and what would be its source (internal or external). Also, how much energy is expected to be generated by hydroelectric power plants, taking into account the El Niño phenomenon, and finally, what is the expected growth year by year in renewable energy sources.
4. The number of probable and proven reserves of non-renewable energy resources and a roadmap of those resources that the country is willing to produce before having to meet its emission reduction requirements.

5. A clear structure of its energy matrix by sector, with clear commitments regarding transition time and resources to finance such transition.
6. The structure of the country in relation to energy demand and energy sources. Because, in many cases, renewable energies are the only reasonable possibility to supply energy services to remote communities, improving their quality of life and facilitating their economic development through productive projects.

If a country is not clear about the 6 variables listed above, it will be swimming in a sea of uncertainty until achieving any energy transition. This is because it will be unable to economically and physically guarantee a constant energy flow for its entire population. Furthermore, it will put at risk important company resources and state institutions that currently grant concessions to private parties for energy resource exploitation and contracts for the generation of renewable energies with firm purchase obligations.

A clear example of a lack of planning is the granting of a 20-year concession for the exploration and exploitation of coal, knowing full well that by year 10, the country's commitment to producing and using coal should be close to 0. In this example, the contracting entity would terminate the concession before the contractually stipulated term. There is also a strong possibility that the business owner to whom the concession had been granted may sue the contracting institution or even the Colombian State, probably with a high chance of success.

Additionally, a country that does not have a clear timetable for its energy transition will be less attractive to investors, since they need to know for sure for how long (in years) they will be able to exploit the resource and the extent that this determines the return on their investments.

13.3 Conflict Arising from the Energy Transition

Bearing in mind the premise indicated at the beginning of this chapter, a country's lack of a long-term plan that efficiently develops and implements a schedule for its energy transition, takes into account its resources, environmental commitments, financing possibilities and expected energy demand, may lead to multiple disputes in the following years. Furthermore, this will be framed not only in purely contractual issues, but also in issues of bilateral or multilateral commitments with a high environmental component.

It is increasingly common for contracts for the construction of large power infrastructure works to contain 'good' or 'best practice' clauses, and that the laws applicable to the contract are not any particular national law, but rather 'public international law' or the 'general principles of international law'.[11] This leads the contracting parties to be subject to principles increasingly aimed at protecting the environment and the restrictive criteria of interpretation in environmental matters. These increasingly common contracts contain international arbitration clauses, and when controversies arises from the performance of the contract, they would be subject to international arbitration.

In fact, by 2018, the Permanent Court of Arbitration of the International Chamber of Commerce of Paris was managing 17 commercial arbitrations related to environmental and energy issues in private commercial contracts. Among these, the following can be highlighted: (i) Ecuador TLC S.A. (Ecuador), Cayman International Exploration Company S.A. (Panama) and Teikoku Oil Ecuador (Cayman Islands) v. the Republic of Ecuador, Secretaría de Hidrocarburos del Ecuador and Empresa Pública de Hidrocarburos del Ecuador (Petroecuador) PCA Case No. 2014–32; (ii) Maynilad Water Services, Inc. (Philippines) v. Republic of the Philippines PCA Case No. 2015–37 and (iii) The

[11] Usually contained in the International Principles of Commercial Contracts (UNIDROIT of the year 2016).

Republic of Croatia v. MOL Hungarian Oil and Gas PLC, PCA Case No. 2014–15.[12]

For its part, as regards International Investment Treaties, the Paris Agreement is expressly mentioned in some of these treaties and the obligations and commitments of each country can become extended obligations of the International Investment Treaty. For example, in the Model Bilateral Investment Treaty of The Netherlands of 2018 private investors must comply with the international environmental commitments of said Agreement therefore, both state and private parties acquire additional commitments alongside the typical contractual commitments related to climate change and GHG emissions reduction.[13]

Finally, a clear example of the conflicts to which a Latin American State may find itself in due to the uncertainty of its energy transition plan, is the multiple lawsuits filed by foreign investors in Spain due to legislative changes that involved cuts in subsidies to renewable energy projects. Such lawsuits were covered by the Energy Charter Treaty,[14] and the majority of these proceedings were decided in favour of the investors.

13.4 Conclusions

The increasingly frequent inclusion of references to supranational rules related to the reduction of climate damaging GHG emissions, and the inclusion of obligations derived from commitments framed in the Paris Agreement by the states in their International Investment Treaties, will lead to an increasing number of commercial and investment arbitration disputes related to the ongoing and evolving energy transition. It is important that each country identifies these risks so that they can react in time and take precautionary measures to reduce any resulting conflicts. Likewise, private investors must consider these contractual clauses and

[12] International Chamber of Commerce, COMMISSION REPORT Resolving Climate Change Related Disputes through Arbitration and ADR ICC publication 999 ENG ISBN: 978–92-842–0554-7 (2019), https://iccwbo.org/content/uploads/sites/3/2019/11/icc-arbitration-adr-commission-report-on-resolving-climate-change-related-disputes-english-version.pdf, p. 53.

[13] Ibid 39.

[14] Multilateral investment agreement executed by countries in Europe and Asia in 1994.

the emission reduction requirements that each country is subject to. This has to be achieved for investors to be clear about the commitment in terms of the GHG emissions reductions of each country, the schedule for compliance with such commitments, and the express and tacit contractual obligations by which they are bound in each case. Doing so means that there will be no future surprises in the contract's performance frameworks.

Part IV

Epilogue

14

Epilogue: Changing Latin America Through the Energy Transformation?

Ignacio Herrera Anchustegui

14.1 Energy Transformation in a Pandemic Context

Can the energy transformation change Latin America? This is the underlying issue that one keeps wondering about while reading the contributions in ' *From Fossil Fuels to Low Carbon Energy Transition: New Regulatory Trends in Latin America*'.

Without the current global-pandemic context, this issue might be natural only to a handful of people, particularly those of us interested in legal and socio-political energy research. However, COVID-19 has accelerated the need to question this. Latin America is one of the planet's most affected regions by the disease, both in terms of health issues and economic downturn. The International Monetary Fund reported in

I. Herrera Anchustegui (✉)
Faculty of Law, University of Bergen, Bergen, Norway
e-mail: Ignacio.Herrera-Anchustegui@uib.no

G. Wood and J. F. Neira-Castro, *From Fossil Fuels to Low Carbon Energy Transition*, Energy, Climate and the Environment,
https://doi.org/10.1007/978-3-031-00299-1_14

2020 that Latin America and the Caribbean had seen its GDP decrease by (-7.0%), more than any other region or group of countries in the world, only behind very hurt economies such as Spain (-10.8%), the United Kingdom (-9.8%), or India (-7.3%). Recovery seems to be on its way but slower than expected. Prospects for 2021 indicated a growth in the region to be around 6.3%, which would still be a negative growth pre-pandemic. 2022 looks worse, with only 3.0% growth.[1]

The pandemic has put the energy transition at the top of the Latin American political agenda.[2] A renewable and sustainable energy transformation combined with Latin America's decarbonization and electrification may spearhead the recovery while also creating opportunities for new business, employment, and sustainable growth.[3] Such a thought is not revolutionary in a global context. The European 'Green Deal'[4] and the 'Fit for 55'[5] proposals are anchored in the same logic: public and private spending ought to be steered towards achieving climate neutrality and fostering green growth.

This energy transformation discussion is partly possible due to the socio-economic progress in Latin America over the last decades. Economic development has improved living conditions and made energy access widespread. In 2019, about 98% of its more than 640 million inhabitants had access to electricity,[6] at relatively low prices thanks in

[1] International Monetary Fund, *World Economic Outlook, October 2021: Recovering During a Pandemic* (Washington D.C., 2021), p. 5.

[2] Banco Interamericano de Desarrollo, *El papel de la transición energética en la recuperación sostenible de América Latina y el Caribe*, Nota Técnica No IDB-TN-02142, April 2021.

[3] See, for example, Carlos De León, Carlos, 'La Transición Energética Latinoamericana Pospandémica', Observatorio Económico Latinoamericano, No. 26 (2021); Banco Interamericano de Desarrollo (n 2); United Nations Economic Commission for Latin America and the Caribbean (CEPAL), *América Latina y el Caribe tiene todas las condiciones para convertirse en un hub de energía renovable con gran potencial en hidrógeno verde*, https://www.cepal.org/es/noticias/america-latina-caribe-tiene-todas-condiciones-convertirse-un-hub-energia-renovable-gran (22 June 2021).

[4] European Commission, *The European Green Deal*, COM (2019) 640 final (11 December 2019).

[5] European Commission, *'Fit for 55': delivering the EU's 2030 Climate Target on the way to climate neutrality*, COM (2021) 550 final (14 July 2021).

[6] The World Bank, 'Access to electricity (% of population)—Latin America (1992–2019)', https://data.worldbank.org/indicator/EG.ELC.ACCS.ZS?locations=ZJ.

part to the energy subsidies (not only for electricity but also for hydrocarbons) in the region.[7] Latin America will increase its energy utilization while also seeking energy efficiency, low-carbon emission solutions, and well-functioning energy markets. However, the energy transition and the re-organization of energy markets may likely negatively impact prices, at least in the short term, derived from the phase-out of energy subsidies, carbon pricing adjustments, and associated costs to new investments in renewable energy. The energy transition is likely to stir social conflict.

This is the context in which Latin America, a young, vibrant, energy-rich, yet sometimes and paradoxically (energy) poor region, contemplates its energy future. Threats posed by global warming and climate change beg radical changes from and for all of us around the world. Blessed with an abundance of natural resources, including hydrocarbons,[8] as well the potential for renewable energy,[9] Latin America finds itself in between different energy sources and different incentives as drivers, as pointed out by *Wood* (Chapter 1). For instance, various Latin American countries have historically depended on their hydrocarbon riches. Venezuela, Brazil, Mexico, and Colombia are illustrations of vigorous activity in the petroleum industry with significant swings between public ownership and privatization. The contributions by *Mata-García* concerning Venezuela's hydrocarbons regulation (Chapter 5) or *Beltrán Nishizaki* related to the energy transition and hydrocarbon dependence in Mexico (Chapter 6) are good examples of these clashes. Is a Latin American renewable energy era truly on its way?

In this *epilogue,* as a Latin American living in a very contrasting Scandinavia, also rich in energy resources, both hydrocarbons and renewables,

---

[7] Milton Fernando Montoya; Ignacio Herrera Anchustegui; et alia, 'Meanings of energy poverty in the South American context: a regional overview' in Íñigo Del Guayo; Lee Godden; Donald D Zillman; Milton Fernando Montoya; José Juan González (eds), *Energy Justice and Energy Law* (Oxford University Press, 2020); Banco Interamericano de Desarrollo (n 2), pp. 20–21.

[8] Venezuela, for example, has the largest proven oil and gas reserves in the world, estimated at 303 billion oil barrels, about 19% of the world's total: U.S. Energy Information Administration, 'Venezuela', https://www.eia.gov/international/analysis/country/VEN.

[9] The International Renewable Energy Agency reports that around a quarter of the total energy consumption in Latin America comes from renewable sources, more than double the world's average: International Renewable Energy Agency (IRENA), 'Latin America and the Caribbean', https://www.irena.org/lac.

and at the forefront of the energy transition globally,[10] I would like to discuss three issues that represent challenges and opportunities for the energy transition in Latin America. Much like the contribution by *Ferney Moreno* and *Aguilar Abuanza*, these reflections seek to understand energy law and energy governance in a broad context, not only anchored on sectoral and technical regulation (Chapter 2). These issues are: resource abundance; renewable energy investment; and renewable-driven energy justice. I discuss thems in the following sections, including some final words related to the energy transformation as an enabler of regional change.

14.2 Energy and Mineral Riches, Between Affluence and Dependence

A recurring theme in the Latin American energy discussion is that of its natural riches. Historically, the debate has mainly been focused on hydrocarbons, as remarked above. Not only some Latin American countries export plenty of these resources, but all states in the region use them internally. In 2016, oil represented 46% of the region's primary energy supply, while natural gas represented about 23% of the region's energy demand.[11] Both are mainly used for transport and electricity generation. While some states are hydrocarbon-rich, most countries in the region are net importers of oil and gas,[12] much needed to fuel their economies, in particular, the transport and industrial sectors.

However, Latin America also has abundant renewable energy sources,[13] including hydropower, sun and wind, existing installed

---

10 For some reflections regarding energy transition, electrification and the interplay between hydrocarbons and its own transition, see: Ignacio Herrera Anchustegui and Aleksander Glapiak, 'Wind of Change: A Scandinavian Perspective on Energy Transition and the 'Greenification' of the Oil and Gas Sector' (Forthcoming 2022). Available at SSRN, https://ssrn.com/abstract=3829455.

11 International Renewable Energy Agency (IRENA), *Renewable Energy Markets Analysis—Latin America* (2016), p. 6.

12 Ibid, p. 10.

13 Ibid, p. 6.

capacity, and perhaps even surprisingly, successful and long experience connected to (renewable) energy auctions as well, ever since the mid-2000s with Brazil and Uruguay as pioneers.[14] Additionally, unlike hydrocarbons, these resources are vastly available to almost all Latin American states. Thirteen countries in the region produce more than 50% of their electricity from renewable energy. Paraguay is the leading example, with 99% of its electricity production being fossil-free based on state-owned hydropower plants, but Costa Rica and Brazil are not far behind with 87% and 84%, respectively.

While traditionally anchored in hydropower electricity generation, concerns related to energy security posed by climate change and hydrological cycles[15] have led states and investors to diversify into wind and solar.[16] Chile as an example of this renewables diversification policy is discussed by *Zamorano* (Chapter 4). Non-hydropower renewable projects have expanded due to solid governance, financial schemes, and some of the world's highest levels of renewable energy availability, such as the case of solar power as well as onshore wind.[17] This combination has made renewable energy production effective and economical, reaching prices as low as US$18 per MWh in Brazil.[18] In the early 2020s, the expansion of renewable energy generation seems to continue to be on the rise, thanks to the abundance of resources as well as the increase of energy use and electrification of sectors.[19]

---

[14] See for more on this: Lisa Viscidi and Ariel Yépez-García, *Clean energy auctions in Latin America,* (Inter-American Development Bank 2020).

[15] Izei Atxalandabaso, 'Renewable energy in Latin America: 5 renewable energy trends emerging from south of Rio Grande' (Rated Power, 2021), https://ratedpower.com/blog/renewable-energy-latin-america/.

[16] IRENA (n 11), p. 10.

[17] Atxalandabaso (n 15).

[18] Rystad Energy, 'Latin America's renewable energy capacity set to skyrocket to 123 GW by 2025' (8 September 2020), https://www.rystadenergy.com/newsevents/news/press-releases/latin-americas-renewable-energy-capacity-set-to-skyrocket-to-123-gw-by-2025/.

[19] The electrification of the transport sector is a pending global issue. There are examples of national programmes introducing, for example, electric buses thanks to international cooperation. See, for example: DW, 'Transición energética: América Latina en la carrera' (12 April 2019), https://www.dw.com/es/transici%C3%B3n-energ%C3%A9tica-am%C3%A9rica-latina-en-la-carrera/a-48310900.

The energy transition, however, also depends on non-renewable resources. This book also brings forth the importance and conflicts around Latin America's mineral deposits. As remarked by *Heredia, Martínez*, and *Surraco Urtubey* (Chapter 8), the energy transition increases the need for the utilization and extraction of minerals, such as the case of lithium, used by batteries. Mining activities for 'new' minerals coexist with a long tradition and strong sector in more traditional mineral-extracting activities related to gold, silver, copper, lead, iron, and coal, as discussed by *Velarde* regarding the mining reform and transition in Peru (Chapter 9).

An abundance of renewable energy resources and minerals needed for their deployment would seem to be positive for the energy transformation and development. Yet, being energy-rich has been a boon and bane for the region. The most infamous illustrative example is my native Venezuela, entirely dependent on hydrocarbon exports,[20] and well discussed in this book. Other countries have been and still are largely reliant on their exports of hydrocarbons and minerals. Mexico (producing about 1.9 million barrels a day of petroleum),[21] Brazil (producing about 2.9 million barrels a day),[22] Colombia (producing about 800.000 barrels a day and declining),[23] or Bolivia (being its three most important export products gas, gold, and zinc, and also the world's largest exporter of tungsten).[24]

Hydrocarbons and minerals are easily exported as these goods can be extracted and transported through ships, for example, but also cross-border pipelines—less common in Latin America than, for example, Europe. However, renewable energy (in the form of electricity) is not easy to transport over long distances as it is lost as heat. This means that

---

20 According to Venezuela's site at the Organization of the Petroleum Exporting Countries (OPEC), Venezuela's oil revenues account for about 99% of its export earnings. OPEC, 'Venezuela', https://www.opec.org/opec_web/en/about_us/171.htm.

21 U.S. Energy Information Administration, 'Mexico', https://www.eia.gov/international/analysis/country/MEX.

22 CEIC Data, 'Brazil Crude Oil: Production', https://www.ceicdata.com/en/indicator/brazil/crude-oil-production.

23 CEIC Data, 'Colombia Crude Oil: Production, https://www.ceicdata.com/en/indicator/colombia/crude-oil-production.

24 OEC, 'Bolivia', https://oec.world/en/profile/country/bol.

while renewable energy can be the engine for the region, it might not be a source of significant exports outside the region, or not to the level hydrocarbons and minerals have been in terms of GDP for many Latin American countries. Furthermore, the abundance of resources creates mixed incentives as the world is still reliant on oil, gas, and coal.[25] Due to existing infrastructure, traditions, and the high prices of these commodities, it is likely that many Latin American countries will keep seeing non-renewable resources as attractive commodities to export.

Can the hydrocarbon and minerals legacy be left behind to create new business opportunities, such as green hydrogen,[26] around renewables? Will the renewable energy abundance diminish the importance of extracting industries and be the force for new industrial policies? May Latin America be able to decouple renewable energy sources from the hydrocarbon legacy and use these to stop depending on oil and gas for its income?[27]

14.3 Funding the Energy Transition, Private Investment, and the Role of the State

Dependence on hydrocarbons and the extraction of natural resources (going back to colonial times with coffee, cacao, gold, and silver) has shaped how ownership (public and private) and investment (foreign and local) have been implemented in energy projects in Latin America. Our countries have had waves of privatization, nationalization, liberalization, expropriation, and renationalization of natural resources and industries.

[25] Together, even after the sharp decrease caused by the COVID-19 pandemic, they represent around 83% of the globe's total energy demand. For some data on this, see BP, *Statistical Review of World Energy* (2021, 70th Edition), p. 12.

[26] CEPAL (n 3).

[27] Geoffrey Wood, 'Managing the Decline of Fossil Fuels: A Long Goodbye?', in: Geoffrey Wood; Keith Baker (eds), *The Palgrave Handbook of Managing Fossil Fuels and Energy Transitions* (Palgrave Macmillan, Cham, 2020) p. 614.

This has been the case concerning oil and gas,[28] even leading to the creation of a failed hydrocarbon-based cryptocurrency.[29]

This book has described a different direction in the case of renewable energy policies, though. Its chapters show Latin America's success and stability compared to its hydrocarbon's past and a swing towards a competition-oriented approach when supporting renewable energy. Brazil, Uruguay, Colombia, and Peru are examples of this. In this path, the role of the Latin American states has been critical. Political commitment, governance instruments and regulations (such as financial incentives and energy auctions), stability, and transparency have enabled 'the further development of market-based financing scheme for renewables'.[30]

This market-oriented approach based on enabling policies and combining private and public funding schemes has worked well, promoting rapid and efficient development of renewable energy projects. The expansion has been fast, totalling 49 GW of installed renewable energy capacity in 2020 from less than 6 GW in 2000,[31]and expected to increase almost threefold to 123 GW by 2025.[32] Despite this success, challenges and conflicts exist, as I discuss below, particularly concerning issues of social justice.

Energy investments are mostly foreign and private. The majority of the top 10 companies with the largest installed non-hydropower renewable energy capacity in the region are European: Enel, Iberdrola, EDF, and Engie.[33] China is also making significant investments in Latin American markets, for both renewables and oil and gas, with more than 50

[28] See for more: Ruben Berrios; Andrae Marak; Scott Morgenstern, 'Explaining hydrocarbon nationalization in Latin America: Economics and political ideology', Review of International Political Economy 18, no. 5 (2011); Osmel Manzano; Francisco Monaldi; Federico Sturzenegger, 'The political economy of oil production in Latin America [with comments]', Economía 9, no. 1 (2008).

[29] Ignacio Herrera Anchustegui and Tina Soliman Hunter, 'Oil as Currency: Venezuela's Petro, a New "Oil Pattern"?'Oil Gas and Energy Law (OGEL) 2 (2019).

[30] IRENA (n 11), p. 11.

[31] Ibid, p. 8.

[32] Rystad Energy, 'Latin America's renewable energy capacity set to skyrocket to 123 GW by 2025' (8 September 2020), https://www.rystadenergy.com/newsevents/news/press-releases/latin-americas-renewable-energy-capacity-set-to-skyrocket-to-123-gw-by-2025/.

[33] Ibid.

projects currently under construction and a value of nearly US$10 billion in Central and South America.[34]

This does not mean that the (foreign) private sector has been the sole investor in Latin American renewable energy. Public entities, particularly national development banks, have historically accounted for a large portion of the investment, up to 1/3 of the total investment pre-pandemic.[35] However, public and local investment lags behind due to the limited financial capacity of Latin American states, more so due to the economic crisis caused by COVID-19,[36] the high cost of capital, and the still limited size and economic capacity of local players, particularly in the supply chain.[37]

Competitive renewable energy projects and markets with public and private participation may coexist. Argentina setting objectives and policies that combine private and public initiatives, with its ups and downs also caused by the pandemic, is a good example of this mix between private and public investment and the state's role, as discussed by *Lanardonne* and *Mazzochi* (Chapter 7).

The market-oriented approach to renewable energy projects is not an isolated trend in Latin American energy policy. During the last decades, governments have taken steps towards the liberalization of the electricity markets or to promote active and responsive consumers. Several countries in the region, such as Brazil, Chile, Colombia, and Mexico, opted to convert their national monopolies into competitive markets with private entities being part of them already in the early 2000s[38]—sometimes seeking efficiency and divestment of public assets, others simply due to fiscal necessities to fund public spending. These liberalized markets coexist in the region with other systems in which publicly controlled entities have a preferential or monopolistic position, such as the case of

[34] The Dialogue, 'China's Renewable Energy Investment in Latin America' (8 September 2021), https://www.thedialogue.org/blogs/2021/09/chinas-renewable-energy-investment-in-latin-america/.

[35] IRENA (n 11), p. 9.

[36] Banco Interamericano de Desarrollo (n 2).

[37] IRENA (n 11), p. 12–13.

[38] Organization of American States, *Policy Reform for Sustainable Energy in Latin America and the Caribbean*, Policy Series No 5, (2004), p. 2.

Uruguay, Costa Rica, Honduras, or Venezuela.[39] Liberalization, competition, and openness to international investment have also spread to the hydrocarbon sector, with the examples of Colombia and Mexico at the forefront.[40]

However, market liberalization and the promotion of energy investment may lead to problems. This book touches upon two of these issues. First, a push for investment and development of renewable energy may lead to poor utilization of public resources and adverse effects on energy consumers. This is a point highlighted by *Venero Carrasco* when studying the Peruvian case (Chapter 3). As stressed by this author, strategies promoting private and public investment in renewables must be anchored on sound regulatory, economic, and holistic energetic frameworks—not in increasing renewables 'just because they are renewables'. Additionally, international investment is often tied to litigation in the form of arbitration. States depend on foreign funds to develop activity and investors seek to obtain profitability. When conflicts arise, parties resort to a neutral forum and neutral rules as studied in the Peruvian case by *Meneses* (Chapter 12).

Can the regulatory stability found while promoting renewable energy markets be used as an example in other areas needing societal change? May Latin America utilize this energy transformation boom to be less dependent on foreign investment through a mix of public and private local investment? Will the market-oriented approach to renewables last and be used to maximize societal welfare?

---

39 Norton Rose Fullbright, 'Renewable energy in Latin America: Central America' (October 2016), https://www.nortonrosefulbright.com/en/knowledge/publications/1e7b0a75/renewable-energy-in-latin-america-central-america; IRENA (n 11), p. 12.

40 For some literature on this see, inter alia, Alicia Puyana Mutis, 'La economía petrolera en un mercado politizado y global. México y Colombia', FLACSO (Mexico, 2015); Guillermo J. García Sánchez, 'The Mexican petroleum licence of 2013', in Tina Soliman Hunter; Jørn Øyrehagen Sunde; Ernst Nordtveit (eds) *The Character of Petroleum Licences* (Edward Elgar Publishing, 2020); Tina Soliman Hunter; Ignacio Herrera Anchustegui, 'Shale gas regulation in Mexico and Colombia: An uncertain future', in Tina Soliman Hunter; Ignacio Herrera Anchustegui; Penelope Crossley, Penelope; Gloria Alvarez (eds), *Routledge Handbook of Energy Law* (Routledge, 2020).

14.4 Energy Justice and the Social Impact of Renewable Energy

This book has emphasized the complex interplay between developing projects to harness renewable energy or extract natural resources and the people living near renewable energy projects. There is no doubt about the associated economic benefits of these types of activities. Renewable energy projects create jobs, develop technology, and increase welfare. However, there are associated negative consequences to them. Examples are visual impact, noise, destruction of untouched nature, occupation of indigenous territory, decrease in property's value, and even expropriation.

Without social balance, the creation of a sense of fairness and an adequate redistribution of resources, conflict is sure to rise. Latin American experiences, mainly connected to new hydropower plants, illustrate this well.[41] Protests and social uproar have delayed and stopped projects in the region. Sometimes renewable energy projects are affected by general conflict and protest in our at times volatile countries, such as the case of a solar project in Chile in 2019 and 2020.[42] More worrisome are claims of human rights violations connected to renewable energy initiatives.[43] Central America seems to be the most affected part in the region, representing 68% of a total of 501 alleged instances of human rights violations by the Business & Human Rights Resource Centre.[44]

While this book does not focus on human rights violations or renewable energy-related conflicts, it does discuss in-depth the complex interplay between local impact and acceptance of new renewable energy projects, as done by *Fonseca Jaramillo* (Chapter 13). Project developers want to conduct their activity swiftly and profitably. States wish to

---

[41] IRENA (n 11), p. 10.

[42] PV Tech, 'Project briefing: Defying COVID and protests in solar's Latin America hotspot', (31 December 2020), https://www.pv-tech.org/project-briefing-defying-covid-and-protests-in-solars-latin-america-hotspot/.

[43] For some discussion of these issues, see, for example: International Service for Human Rights, *Mexico |What you didn't know about renewable energies: the struggle of the Binnizá people of Unión Hidalgo* (17 June 2021).

[44] Business & Human Rights Resource Centre, 'Renewable energy (in) justice in Latin America' (6 August 2021), https://www.business-humanrights.org/en/from-us/briefings/renewable-energy-injustice-in-latin-america/.

promote investment, economic activity, and they must comply with international obligations towards reducing greenhouse gas emissions and increasing the deployment of energy. Citizens, however, are affected by the projects and sometimes see their well-being hindered while not being (sufficiently or fairly) compensated. Balance is needed to comply with governance, climate, and economic objectives, as well as reduce societal conflict and create a sense of energy justice among citizens.

How to find such an equilibrium and create sufficient grounds for social acceptance for renewable energy and resource extraction projects is studied under the 'social license to operate' concept. As *Vescovi Chedid, Pereira, dos Santos*, and *Medeiros Costa* put forth, 'social licence to operate' describes the approaches undertaken by project developers to attain an informal authorization from the local community in which they are established to conduct their activity without conflict (Chapter 10). When studying the Colombian case *Bustos, Zapata Sánchez* and *Bastidas Reyes* identify two key elements needed to improve social acceptance of projects: citizen participation (through prior consultations and local communities), and land restitution or monetary compensation (Chapter 11).

Will Latin America find a robust balance between project developers' interests, local communities' needs, and national objectives to increase renewable energy generation? Can the development of renewable energy and derived economic growth be used to create a sense of fairness and break the pattern of state-dependence in which many Latin Americans live? How will the rise of renewables affect the granting of energy subsidies? Will it lead to its phase-out and substitution by other more efficient redistribution mechanisms?

14.5 Concluding Words: Energy Transformation as the Promotor of Regional Change

From Fossil Fuels to Low Carbon Energy Transition: New Regulatory Trends in Latin America may not answer whether the energy transformation will change Latin America. However, the contributions in this book bring forth pivotal topics related to future regional and national energy policies. Doing so will continue stirring the debate of social change thanks to or because of the energy transformation. I would like to close this contribution with what I hope will become three derived effects of the energy transformation in Latin America.

First, the regional energy transition in Latin America could lead to more and tighter regional coordination, and it would undoubtedly benefit from it.[45] Despite historical, cultural, and linguistic ties, our region is still behind other examples of integration, such as Europe but also parts of Asia. While regional cooperation agreements, such as Mercosur, the Andean Community of Nations, the Caribbean Community, or the Community of Latin American and the Caribbean States, have evolved over the years and promoted (energy) integration, including cross-border projects, we are far from a common energy market. Perhaps this would not be entirely possible in our region due to geographic and political complexities, or maybe those integrated markets might have to be more local. However, coordinated policies, experience sharing and tight dialogue between states, policymakers, and academia is needed. This will facilitate large projects, more cross-border infrastructure, and prevent instances of prisoner's dilemma, such as it could happen in the case of carbon pricing.[46]

Second, developing efficient, sustainable, and socially fair renewable energy projects may promote technological and industrial progress. Such is the vision in Europe related to the 'Green Deal' and green growth. Latin America, blessed with its abundance of renewable natural

---

[45] See also highlighting the importance of integration, Kadri Simson; Thauan Santos, 'Transición energética y áreas geográficas: qué está sucediendo en Europa y América Latina?', Enel Green Power, https://www.enelgreenpower.com/es/learning-hub/debates/transicion-energetica-europa-america-latina; "Regional integration is urgently needed to unlock new sustainable growth opportunities and to improve the resilience of energy systems across the region", World Energy Council, *World Scenarios 2017—Latin America & The Caribbean Energy Scenarios* (2017), p. 2.

[46] Several Latin American countries have forms of pricing carbon, with Mexico, Colombia, Chile, and Argentina as good representatives. The literature points out, however, that national, and not regional schemes may create inequalities as well as lead to increased prices of energy but also almost any other good, to the detriment of the poorest societal sectors. See for a thorough discussion of carbon pricing schemes and policy suggestions: Banco Interamericano de Desarrollo (n 2).

resources, may use this opportunity to create a new industrial policy anchored on technological development in addition to natural resource exploitation. Also, this should be used to promote new business opportunities and a vibrant renewable energy value chain in the region, from the construction of solar panels or wind blades to cables installation, to project management.

Third, the energy transition and the development of renewable energy projects in Latin America may be used by our countries and politicians to achieve two goals. First, to reduce the dependence on the extraction of non-renewable energy resources, such as hydrocarbons and minerals. While historically fundamental for the development of our societies, it has also made us vulnerable and reliant on one resource and industry—with the dangers that lack of diversification bring. Secondly, policy goals and objectives for renewable energy development should include redistributive mechanisms and schemes that properly allocate costs and serve to reduce the social gap and inequality in our region. These models should inspire measures also outside the energy realm. The Latin American renewable energy transition must be clean and rapid. Perhaps more importantly, it must be inclusive and fair.

Index

G. Wood and J. F. Neira-Castro, *From Fossil Fuels to Low Carbon Energy Transition*, Energy, Climate and the Environment, https://doi.org/10.1007/978-3-031-00299-1

FSC
www.fsc.org
MIX
Papier | Fördert gute Waldnutzung
FSC® C083411

Kolibri 360 GmbH
Ferdinand-Jühlke-Straße 7
99095 Erfurt, Deutschland
produktsicherheit@kolibri360.de